GREEN HOME

PLANNING
AND BUILDING
THE
ENVIRONMENTALLY
ADVANCED HOUSE

WAYNE GRADY

CAMDEN HOUSE

Canadian Cataloguing in Publication Data

Grady, Wayne
   Greenhome : planning and building the environmentally advanced house

Includes index.
ISBN 0-921820-69-0

1. Dwellings - Ontario - Waterloo - Environmental engineering. 2. House construction - Environmental aspects - Ontario - Waterloo. I. Title.

TH4811.G73 1993   690'.8370472   C93-094506-9

Published by Camden House Publishing
(a division of Telemedia Communications Inc.)

Camden House Publishing
7 Queen Victoria Road
Camden East, Ontario K0K 1J0

Camden House Publishing
Box 766
Buffalo, New York 14240-0766

Trade distribution by
Firefly Books
250 Sparks Avenue
Willowdale, Ontario
Canada M2H 2S4

Box 1325
Ellicott Station
Buffalo, New York 14205

Design by
Linda J. Menyes

Illustrations by
Ian Grainge

Filmwork by
Hadwen Graphics Limited, Ottawa, Ontario

Printed and bound in Canada by
Metrolitho Inc., Sherbrooke, Quebec

CONTENTS

THE SURVEY 5

GROUND WORK 35

THE FOUNDATION 65

FRAME WORK 91

THE MECHANICALS 125

FINISHINGS 163

OPENING 193

SOURCES 199

INDEX 205

FOR MORGAN AND CLAIRE—THE FUTURE

□ □

A GREAT MANY PEOPLE CONTRIBUTED TO THE CONSTRUC-
TION OF THE GREENHOME, AND SO TO THE WRITING OF
THIS BOOK. STEVE CARPENTER OF ENERMODAL ENGINEER-
ING INITIATED THE GREENHOME PROJECT, AND HE AND
JOHN KOKKO DEVOTED MANY HOURS EXPLAINING IT TO
ME. I WOULD ALSO LIKE TO THANK THE OTHER MEMBERS
OF THE DESIGN TEAM, ESPECIALLY ELIZABETH WHITE,
TONY KRIMMER, IAN COOK, WERNER REITER, RICHARD
REICHARD AND ED BORDEAU.

TIM MAYO OF ENERGY, MINES AND RESOURCES WAS VERY
HELPFUL, AND OLIVER DRERUP TAUGHT ME MUCH ABOUT
LOW-ENERGY BUILDING AS WELL AS GRACIOUSLY OFFERING
TO READ THE FINISHED MANUSCRIPT.

AT CAMDEN HOUSE, I'D LIKE TO THANK EDITOR TRACY
READ FOR HER PATIENCE, ART DIRECTOR LINDA MENYES FOR
HER BRILLIANCE, ASSOCIATE EDITOR MARY PATTON FOR HER
SKILL AND ILLUSTRATOR IAN GRAINGE FOR HIS TALENT.
THANKS ALSO TO EDITORIAL PRODUCTION CONSULTANT
SUSAN DICKINSON AND ASSOCIATES CATHERINE DELURY
AND CHRISTINE KULYK.

MOST OF ALL, I WANT TO THANK MERILYN SIMONDS MOHR,
WHO, IN EDITING THIS BOOK, BROUGHT TO IT A DECADE OF
HER EXPERIENCE AS THE ENERGY AND SHELTER EDITOR OF
HARROWSMITH MAGAZINE.

THE SURVEY

□

Lot 92 didn't look like much when I first saw it in August: an average corner lot in an average subdivision. It was conspicuously empty. The lots on either side already had houses on them. One was a two-storey brick job with a detached double garage, a driveway of loose red gravel and a Japanese garden in the backyard. A sign on its front lawn proclaimed it The Rotary Dream Home; the local Rotary Club had bought the house from Arnold Freure, the subdivision's developer, for $200,000 and was raffling it off as a fund-raiser. Tickets were $100. Even as I stood there, a young couple in a white Mazda pulled úp in front of the house and stared blankly out at the brick façade. I tried to see the Dream Home through their eyes. It looked very compact. In fact, it looked tiny. The garage was nearly half as big as the house.

The other house was only slightly bigger than the Dream Home: a two-storey frame house with white vinyl siding and a black asphalt roof. Its lawn had been so hastily laid out that some of the sod was still wrinkled up against the basement wall, like a throw rug in a slippery hall.

Between the two houses, Lot 92 was a promise waiting to be fulfilled. It had been raining on and off for weeks – this was the summer of Pinatubo, the coldest, wettest August in 94 years. Scraped clean of its topsoil, the lot was little more than a pool of clay-coloured water ringed by a fresh, white curb. A sign at the centre called it the site of the Waterloo Region Greenhome and provided a list of benefactors and well-wishers – the Kitchener-Waterloo Home Builders Association, the City of Waterloo, Union Gas, Ontario Hydro, Energy, Mines and Resources Canada. To the right of the sign, a young colt of a maple sapling stood shaking the drizzle off its leaves. It was the only tree I could see anywhere. It had been set there a month ago as part of an official ceremony to launch the Greenhome. Before many months would pass, the sapling would be part of the landscape design of the most advanced, the most energy-efficient, the most commonsensical house ever built in Canada.

□ □

The subdivision is called Southwind and is on the outskirts of Waterloo, Ontario. For a long time, Waterloo was known as an insurance

6

town. The Waterloo County Mutual Fire Insurance Company started there in 1863, became the Ontario Mutual Life Assurance Company in 1870 and is now known as the Mutual Group, one of the largest life-assurance companies in North America. Life is assured in Waterloo, but fire is insured. The North Waterloo Farmer's Mutual Fire Insurance Company opened in 1874, the Mercantile Fire Insurance Company in 1875 and the Dominion Life Assurance Company in 1889.

All of them are still there, but since the 1950s, Waterloo has become more a university town. The University of Waterloo opened in 1958, followed by Waterloo Lutheran University (since renamed Wilfrid Laurier University) a few blocks away, and now the city claims to be second only to Toronto as a centre of higher learning. As far as image goes, there isn't a lot to choose between an insurance town and a university town. A quiet, residential place, plenty of trees, an old Carnegie library, one good bookstore, three good restaurants. It is as unlike its beer-hall sister city of Kitchener as two cities can be. While Waterloo was wooing insurance companies and college students, Kitchener was industriously drawing in factories and labourers, and the difference between the two is usually expressed in the colour of the collars their inhabitants wear to work.

There have been sporadic attempts to amalgamate, but they have always been successfully resisted by Waterloo. The latest, in 1970, was settled by public referendum, in which 83 percent of Waterloo's residents voted no. Nonetheless, Kitchener and Waterloo are almost always verbally linked. The exit sign on Highway 401 reads "Kitchener-Waterloo." The newspaper that once proposed that in the event of amalgamation, the name should be changed to Kitchenwater is called the *Kitchener-Waterloo Record*. Kitchener-Waterlooers attend the Kitchener-Waterloo Art Gallery, the Kitchener-Waterloo Chamber Orchestra, the Kitchener-Waterloo Bilingual School, the Kitchener-Waterloo Friendship Group for Seniors, the Kitchener-Waterloo Optimist Club, if they're optimists, or the Kitchener-Waterloo Crisis Pregnancy Centre if they're not.

Max Braithwaite, in his 1974 book *Ontario*, noted that "Kitchener-Waterloo is a German community." The German part, at least, is true. The first settler in what is now the city of Waterloo was a German Mennonite named Abraham Erb who, in 1806, drove up from Lancaster County, Pennsylvania, in a Conestoga wagon and settled on 448 acres of low, wet cedar swampland on the banks of Laurel Creek, which empties its heavily silted water into the Grand River, which in turn debouches into Lake Erie about 30 miles west of the Niagara River. Erb built a gristmill on Laurel Creek, creating a millpond that is now called Silver Lake, 7 and his establishment became a kind of social centre for the district, being the only gristmill between Perth County and Preston. The township was named Waterloo in 1817 in commemoration of the defeat of Napoleon, and Erb named his mill site Waterloo for the sake of consistency and because he believed that the defeat of the French by a predominantly German army was something to commemorate – a feeling that would come back to haunt the community in 1914. He refused to parcel out his land, however, so he had the place pretty much to himself until 1829, when for some reason, he suddenly decided to sell half of it, including his mill, to Joseph C. Snider, who started a distillery.

After that, more settlers began to arrive, some from Pennsylvania and many from Germany; within a year, Joseph's son Daniel had opened the district's first post office; then John Hamilton opened a general store and Henry Bowman began to operate a tavern where, in 1844, he began selling not only Snider's schnapps but also David Kuntz's lager beer. Since then, the city has become almost as well known for brewing and distilling as it is for insurance premiums and midterm exams.

Since 1973, the city of Waterloo has been called the Regional Municipality of Waterloo, a metropolis that incorporates five townships, four towns, three villages and three cities (though not Kitchener) into a single tax shed with a total population of about 75,000. A few years ago, Waterloo acquired an unfair reputation as a bedroom community for Toronto, a mere 45 minutes away on Highway 401 – unfair because the commuter train's last stop is at Guelph, and the people of Waterloo are glad it is. Still, until the recent recession, Waterloo was one of the fastest and most steadily growing urban communities in the country. And even with the recession, as the houses dotting such suburbs as Southwind attest, new-house starts are continuing to hold their own.

The house on Lot 92 won't *look* much different from its neighbours. It, too, will be a frame house with a built-in garage and a sharply peaked roof over a wide front porch; there will be a little deck off the back for weekend barbecues; the yard will be nicely landscaped. It will sell for about the same price as the Dream Home, but the comparison really stops there.

Certainly, nothing about its exterior will tell the casual observer that the Greenhome sits, in fact, at the cutting edge of low-energy house design and construction. But it will consume less than one-fifth the energy gobbled up by the identical-seeming bungalows across the street that were built to the specifications of the National Building Code. The average new house in Canada today uses 50,000 kWh of electricity a year, about half of that for space heating and another tenth for hot water;

in the Waterloo Greenhome, space heating and hot water will use only about 7,500 kWh of electricity, at a cost of about $400 a year. The appliances in the house across the street soak up 20,000 kWh of electricity a year; the Greenhome's will take up 5,000. While the house across the street has 6-inch walls stuffed with pink fibreglass for an insulation value of about R-20, the Waterloo Greenhome's walls will be more than 12 inches thick, with blown-in insulation that will give them an R-value of 40. The house across the street has double-glazed windows with aluminum spacers that look very nice but are only marginally more efficient than the old single-pane sashes in use 25 years ago. The Greenhome's superwindows will be the most efficient on the market, with a midpane insulation value of R-17 — almost as much as the house across the street has in its walls. While it was being built, the house across the street generated 2½ tonnes of construction waste that had to be hauled to a landfill site; no construction waste at all from the Greenhome will go to a landfill. None. Everything will be recycled. And much of the house will be made from recycled material.

In other words, whereas the average house in Canada today costs about $3,500 a year to run, the Waterloo Greenhome will cost $800. If the difference were applied to its mortgage payments, the Greenhome would be mortgage-free 10 years ahead of its neighbours and continue to cost a quarter as much to run for the next 50 years. But money isn't the only factor and in fact isn't even the most important one. The house across the street represents a constant and increasingly insupportable drain on the environment and on the world's dwindling energy resources; the Waterloo Greenhome will, to borrow a phrase applied to architecture by Pierre Trudeau, allow its eventual owners "to live lightly on the land."

□ □ □

The second time I saw Lot 92 was at 7 o'clock on the morning of September 28, and it was much changed. Its contours had been tidied up by a second pass with a bulldozer. It was level, with a slight rise toward the Dream Home's driveway. Sod had been laid along Westvale Drive, between the sidewalk and the curb, and a 4-foot swath of wood chips ran around the lot's perimeter. The morning was cold, but the sun warmed the car nicely as I sat waiting for the surveyors, drinking Tim Hortons coffee and watching the fog lift from the valley behind the subdivision. The streetlights along Westvale went out at 7:08. At 7:22, a small wedge of Canada geese honked overhead, about 30 feet above my car, looking for a place to land. One of the Dream Home's flags snapped in the wind like a gunshot.

I was reading *The Globe and Mail*. There had been a slight upturn in the sale of uranium for nuclear reactors: a Saskatoon company, Cameco Corp., had reported revenues down 10 percent from last year but profits up 40 percent. This was called good news. "Uranium mining does provide jobs," a Toronto analyst was quoted as saying. "The economics have to be looked at very seriously. Ideology may have to take a back seat." In another story, a survey conducted by Environics Research found that Canadians were "suffering from a form of environmental fatigue," that many of us were satisfied we had done enough for the environment and were moving on to other issues. One in four of us, it said, still puts hazardous waste out with the regular garbage and thinks the country already has enough national parks.

I was glad when, at 9 o'clock sharp, a grey van pulled onto Bridgewater Street and parked in front of me. The two men who got out were Brian Vander Vlught and Jerry Smith, who worked for Campbell, Wyman and Auer Ltd., a firm of land surveyors. Jerry took a shovel and a metal detector from the back of the van and marched off to the corner of the lot while Brian strapped on a belt hung with equipment: a hammer, a bunch of red-painted spikes, a roll of fluorescent red surveyor's tape, a long tape measure and a solid brass plumb bob. The belt buckled in front and was held up with a pair of broad red suspenders.

"With any luck," he said, striding across the lot to join Jerry, "there'll be an SIB, that's a standard iron bar, a surveyor's stake, at the corner of the lot, and we can sight on that. We've got to run lines along this street and down that street so we can level and position the house."

Jerry was already sweeping the metal detector over some wood chips a few feet from the sidewalk at the corner. The detector, which looked like a yellow walking stick with a radio at the top end, emitted a low whine until it sensed the iron bar, when it suddenly started to sound like a chain saw having a nervous breakdown. Jerry dug with his shovel and uncovered the SIB, the top 8 inches of which had been bent nearly horizontal by the bulldozer. "Doesn't give us much to go on," said Brian.

The iron bars had been put in place by other surveyors when the subdivision was first laid out. There are thousands of them. Along with a series of cement posts called monuments and another, smaller series of universal transverse Mercator coordinates, or UTMs, the SIBs act like the grid points on a map; they define the boundaries of each separate building site, dictate where the roads and sidewalks will go, where the sewers and storm drains, power lines and telephone cables are to be buried. The survey lines for each house are taken from these stakes. The problem is that after the surveyors meticulously put them in place, graders and bulldozers come in to scrape down the roadbeds, pile the top 6 inches of topsoil into huge mounds to be put back after the houses are built and generally deface the landscape to ready it for total and permanent alteration. They often bend the iron bars out of position in the process.

"We could just guess where the bar was before it was bent over," says Brian. "But they don't give you a lot of slack in positioning these houses. A two-storey house has a minimum side-yard clearance of 1.829 metres, and a one-storey house has 1.2 metres, and they allow only a 2-centimetre tolerance. Two centimetres isn't a whole lot to play with, and if this bar is out by 4 centimetres, then there goes my tolerance. We've talked to developers about putting a different system in place," he says, "a system called deferred monumentation, which means putting the roads and houses in first and then the monuments and bars. Everyone would have to work from some independent control structure."

"You mean a map?"

"Yeah. But it turned out that would cost more than the present system. So here we are."

The SIB at the northwest corner of the lot is gone altogether, so we walk farther along Westvale, past the Dream Home to where, according to Brian's independent control structure, the next one is supposed to be. Jerry digs where the detector screams at him to dig, and there it is, more or less on the Dream Home's front lawn. From there, Brian paces off the distance to the next bar, which is in a more exposed spot on an undeveloped site. We then walk back toward our lot, cross Bridgewater and uncover a third bar on the corner by peeling back a flap of sod. Jerry and Brian now have three unbent SIBs running along Westvale, enough to enable them to reposition the bent bar on the Greenhome lot.

While Brian draws up a site plan on a sheet of grid paper, marking in the correct position of the corner bar, Jerry sets up their transit over the last SIB they uncovered, across Bridgewater.

"We don't really call them transits anymore," he says. "Now they're known as electronic theodolites, total units. This one is a Sikkosha Set 3. You just line it up on a position and press a button, and it sends out a beam of light that hits a crystal, measures the time it takes for the light to come back and gives a digital readout of the distance and the angle." By now, Brian is holding his plumb bob over the second SIB. The plumb bob is suspended from a string, and at the top of the string, in Brian's hand, is a crystal. Jerry sights the theodolite on the crystal, pushes two of the 20 buttons on the side of the unit, waits two seconds, then reads the result.

"Sixty-five point four seven six metres," he shouts to Brian, who writes the number down on his site plan, then walks back to where we're standing. "Okay," he says, "now we've got to work back along this street to see if we can find some bars at the north end of the lot."

"We should have done that first," says Jerry. Brian looks at him, then turns and goes back along Bridgewater. "Brian and I are both crew chiefs, normally," says Jerry, "and we have different ways of doing things. But with the industry so slow these days, we don't have enough work to keep separate crews busy, so we go out and do it ourselves."

I ask him how slow the industry is. Jerry shakes his head. "Two years ago," he says, "we were doing a lot of work for Northlake Homes, the other side of Waterloo. In 1990, Northlake built 200 houses; in the past two years, I don't think they've built more than 20, total. I look around at all these new houses going in here, and I wonder where the hell all these people came from, where are they working?"

There are no SIBs left at the back of the Greenhome lot either, as it turns out, so Brian and Jerry move the theodolite and reestablish a lot line running along Bridgewater. Working from the new corner SIB, they are now able to position the house on the lot. This is done with four wooden stakes, each of which is placed 5 feet from where the walls of the house will be. Repositioning the SIBs has taken 2½ hours; pounding in the wooden stakes takes about 15 minutes. By noon, the van is neatly packed and ready to pull out, and Lot 92, staked and measured, is ready for digging.

□ □ □ □

IT IS DIFFICULT TO MAKE PREDICTIONS, ESPECIALLY ABOUT
THE FUTURE.
—NIELS BOHR, NUCLEAR PHYSICIST

"Whenever I'm interviewed about the Waterloo Greenhome," says Tim Mayo, "I'm always hearing, 'Yeah, but how much is it going to cost?' And I say, 'Look, the Greenhome is a prototype. It cost Thomas Edison a lot more than 69 cents to build the first light bulb.' "

As manager of the Buildings Group, Energy Efficiency Division, at Energy, Mines and Resources Canada (EMR), Tim has been interviewed a lot about the Greenhome; he is, in a sense, paying for half of it. He invented the EMR's Advanced Houses program, which is sponsoring

10 new low-energy houses across Canada, of which the Greenhome is one. The program, announced in April 1991, is Tim's attempt to drive home the point, at last, that energy efficiency in domestic construction is not only possible, not only desirable, but absolutely necessary given the rate at which our current sources of energy are being depleted.

Despite *The Globe and Mail*'s doom-and-gloom report on our environmental ennui, low-energy architecture shouldn't be a hard sell. Other surveys show that worrying about the environment is still high on most people's list of the things that keep them awake at night, far ahead of the price of uranium. A survey conducted in July 1992 showed that 40 percent of Canadians are "true-blue greens" – people who think that green is good and who also act on their concerns and speak out on such issues as pollution and environmental degradation. And 3 percent are "greenback greens," people willing to pay more for environmentally responsible products. A focus group of house buyers questioned by the Ottawa-based environmental group Energy Pathways indicated that some of them are even willing to pay more money for a smaller house that is intelligently designed and energy-efficient, phrases that practically define the Greenhome.

"I think if enough consumers raise these issues with builders," says Tim, "builders will realize that if they respond to them, they can sell houses. The problem is that builders don't have access to information on environmental issues as they relate to building houses. So with the Advanced Houses program, we're focusing on builders as the prime target, but since not enough consumers are asking these questions, we are also trying to make the public more aware of what's going on, what kinds of things are available, or soon will be. Broadly speaking, then, our intent is to encourage innovation in the development of new energy-efficient technologies in the housing industry and to get the message out that energy-efficient houses are just around the corner, so to speak."

There are thus two roads leading to the Greenhome: a short road and a long road. The short road starts in Tim Mayo's office, and we'll follow that one in the next chapter. The long road starts back in October 1973, when the Organization of Arab Petroleum Exporting Countries (OAPEC) announced to the world that they were imposing a six-month freeze on oil exports to North America (and reducing those to Europe by 5 percent) in retaliation for the United States' support of Israel in the Arab-Israeli War.

The reaction to the embargo was panic, but the reasons for the panic were complicated. Throughout the booming 1960s, world energy use had increased by about 6 percent a year; oil's share was also increasing, accounting for 39 percent of commercial energy supplies in 1960 and 51 percent in 1974. The idea of oil shortages did not originate with OAPEC. It was introduced to the North American consciousness as early as 1969, when U.S. Geological Survey scientist King Hubbert predicted that at the current rate of consumption and increase, 80 percent of the world's crude-oil reserves would be exhausted by the year 2032. But Hubbert added that he expected new oil reserves to be discovered, so no one really felt compelled to turn down the thermostat quite then. The following year, however, Canadian environmentalist Donald Chant pointed out that Hubbert's prediction should be read in the light of the American government's stated aim "to triple energy production by the year 2000." Chant also noted that University of California professor Ken Watt had warned that "if the supersonic transport airplane is introduced, the fuel demands of such planes will deplete crude-oil reserves much faster, in all probability by 1990."

The OAPEC embargo drove Chant's point home a little deeper. If Hubbert made us realize that there was a limit to the amount of oil available to us, then OAPEC showed that the limit could be imposed as a result of something as seemingly arbitrary as our sympathies for one side or another in a distant foreign war. The double realization produced an extreme shock to our collective nervous system.

Also, the OAPEC crisis coincided with a sudden gigantic increase in the price of crude oil from all sources – the price of oil from Venezuela increased by 400 percent in one year, and the cost of extracting domestic oil shot up 200 percent, with the result that, from February 1973 to August 1974, the retail price of fuel oil went up 60 percent. Most of us blamed the increase on the embargoes, even though OAPEC accounted for only 35 percent of all the oil used in North America. Those with somewhat longer vision realized that the increasing costs were more directly attributable to decreasing reserves. A new phrase entered the language: nonrenewable resources.

At the end of 1973, proven world reserves of oil were estimated to be about 630 billion barrels. But most of that oil was in – or rather under – the Middle East; reserves available to the entire western hemisphere, including those in the United States, Canada, Latin America and the Caribbean, were only 12 percent of the world's total, or enough to last the United States (if it got all of them) a little over a dozen years at its current rate of consumption of 6.3 billion barrels a year.

A few months after OAPEC, something happened in Alberta that seemed to symbolize the emerging oil crisis. Alberta was the Texas of Canada as far as oil was concerned; its oil patch was so productive that

Alberta residents didn't have to pay any sales tax. That prosperity began in 1947 with the discovery of a major oil field near Leduc, a few miles south of Edmonton. Imperial Leduc No. 1, as that legendary well was called, went dry in February 1974.

No more oil. The idea was a major shift in the way we could think about the future. Oil had been slaking our thirst for industrial growth for less than a century, but our economic prosperity was firmly fixed to it, as if we were addicted to the stuff. Running out of oil was like running out of water or trees or clean air. OAPEC was the first in a series of death knells for our sense of North America as an inexhaustible land of plenty. We may no longer have believed we could place a chicken in every pot, but we clung tenaciously to the idea of a huge furnace in every basement and a big car in every garage. In the United States in 1973, 46 percent of the nation's homes were heated by oil – in Canada, it was over 71 percent, at an average cost of only $130 a year per household. Suddenly, we had to face the fact that either there would be no oil or what there was would be so expensive that most of us wouldn't be able to afford it. Even the most optimistic forecasters were talking about paying as much as $100 a barrel for crude oil that only months before had cost less than $5. That $130 home-heating bill was about to be multiplied by 20.

The long-term solution to the problem was clear: reduce our dependency on oil. In terms of a national energy policy, this meant two major thrusts: conservation – reducing the amount of energy we needed – and switching from oil to other, more stable energy supplies. Conservation meant smaller, more fuel-efficient cars (Detroit's first small cars were introduced in the early 1970s, when someone at General Motors finally noticed how many Volkswagen beetles were in the GM parking lot), lowering the speed limit on highways and putting more insulation in house walls. And switching from oil meant turning to natural gas, which was already being used to heat 27 percent of Canadian houses in 1971 and could be pipelined to the energy-hungry east; the use of electricity for home heating could also be encouraged (in 1974, only 4 percent of all houses were electrically heated; in most studies, electricity as a heat source was listed as "Other," along with coal oil and sawdust).

A report released by the federal government in 1973, called *An Energy Policy for Canada*, outlined residential energy use in 1969 and projected it to the year 2000. The forecast included four energy sources – coal, oil, gas and electricity – and predicted that by the turn of the century, all new houses would be heated either by gas or by nuclear-generated electricity and that oil-heated houses would fall by the wayside through at-

15

trition. Assuming that Canada would have 10 million houses to heat in the year 2000, the report estimated that 6.1 million of them would be heated by gas, 1.2 million by electricity, none of them by coal and 2.7 million (those built before 1969) by oil.

It's an interesting study. A review of it published in *Habitat* magazine in June 1973 summarized it this way: "Basically, it appears that the energy problems of Canada, which have been precipitated by the present 'crisis,' have arisen because a growing nation has been consuming oil and natural gas at low and stable prices, based on cheap and accessible reserves that are being rapidly diminished. To rectify this problem, it is essential that prices for all types of energy should rise to encourage exploration and development of further reserves, and economy in the use of fuel must be practised up to the year 2000, after which a change to an 'electric' society based on nuclear energy is expected." In other words, the government had published what it hoped would be a self-fulfilling prophecy: higher prices for oil and gas, with an eventual switch to nuclear energy. The report made no mention at all of renewable energy sources; the word "solar" did not appear anywhere in the discussion.

□ □ □ □ □

NOW, IN HOUSES WITH A SOUTH ASPECT, THE SUN'S RAYS
PENETRATE INTO THE PORTICOES IN WINTER, BUT IN
SUMMER, THE PATH OF THE SUN IS RIGHT ABOVE THE ROOF
SO THAT THERE IS SHADE. IF, THEN, THIS IS THE BEST
ARRANGEMENT, WE SHOULD BUILD THE SOUTH SIDE LOFTIER
TO GET THE WINTER SUN, AND THE NORTH SIDE LOWER TO
KEEP OUT THE COLD WINDS.
—SOCRATES, 360 B.C.

It is one of life's delicious ironies that solar energy is in fact a by-product of nuclear energy. Light and heat from the sun are created by the fusion of hydrogen into helium. Normally, hydrogen nuclei repel each other, because each consists of a single positively charged proton. At the centre of the sun, however, where temperatures can reach up to 20 million degrees Kelvin and hydrogen atoms are packed so tightly that their density is 12 times that of lead, these single protons frequently combine to form nuclei with two protons, which is helium. In a single second, the sun converts 594 million tons of hydrogen into 590 million tons of helium. Since, according to Newton's law of constant composition, matter can be neither created nor destroyed but merely changed in form, the 4 million tons of matter lost during the transformation of hydrogen into helium are converted into pure energy. A lot of pure energy.

Only a tiny fraction of that energy reaches Earth, but it is the equivalent of 178 billion kilowatts of electricity a day. About 40 percent of this energy is radiated back out into the atmosphere. Another 50 percent is absorbed into the earth, or whatever else it strikes, as heat. The remaining 10 percent is available for conversion into electricity. Even at that, a single square metre of it would produce 930 watts of electricity if we could only find some way to convert all of it.

For the longest time, we had no way of converting any of it. Then, in 1839, French physicist Alexandre-Edmond Becquerel was fooling around with a battery consisting of an anode and a cathode in a conductive solution, and he noticed that it produced more electricity when exposed to sunlight than when it was not. He didn't know why this was so, and he didn't have any use for the result, but being a good scientist, he made a note of it just the same. Later on, in 1889, an American physicist named Charles Fritts covered a strip of selenium with gold film and exposed it to sunlight. Like Becquerel, he didn't know what caused an electric current to run from it, but he knew what to do with the result: seven years earlier, Thomas Edison had invented the light bulb.

What Fritts had discovered was the photovoltaic cell, a method of converting the sun's energy into electricity, which would become interesting as a substitute for oil in home heating if we could convert more than 10 percent of the photons that bombard our planet daily into electrons. But we can't, at least not yet. Of more use to us in our post-OAPEC darkness is the 50 percent we *can* harness in the form of heat. The Greenhome will have a small photovoltaic cell on its roof, but the limited amount of electricity it produces will be used to run another piece of solar equipment: a hot-water heater. It has taken us a long time to figure it out, but harnessing the sun's energy as heat is a lot easier than using it to make electricity.

There are two ways to make heat with the sun's energy. One is to lie on a beach and let the sun's rays warm your body; this is called passive solar and obviously doesn't require a lot of technology. Devising some mechanical method of storing that warmth and moving it to a different place—fanning heat from a hot rock into your tent, for example—is called active solar. In the years following OAPEC, many of the first low-energy houses built in Canada used active-solar technology for space heating. And to be fair, despite the predictions of the government's official energy policy, some federal and provincial departments were beginning to endorse active-solar development. Doug Lorriman's house, for instance, built in Toronto in 1975 and designed so that most of its space-heating would come from an active-solar system, was made possible by a $60,000 grant from the Ontario Ministry of Energy. Lorriman installed 690 square feet of glass on the south-facing wall of the house; under the glass were coils of copper tubing through which water flowed, to be heated by the sun and then transferred through two 2,500-gallon storage tanks in the basement, where it gave up its heat to the cooler water in the tanks, then moved back up to the tubing to collect more. When the temperature of the water in the tanks reached 115 degrees F, it was circulated through more tubing installed inside the house's ductwork; air blown through the ducts was heated by the water in the tubing. This system, Lorriman predicted, would supply as much as 70 percent of the house's heating requirements in winter.

As indicated by the $60,000 government contribution to Lorriman's house – almost half of the total construction cost – active solar was an expensive proposition, nearly as far out of reach to the average homeowner as oil was expected to be. Solar One, the first fully solar-powered house built in the United States, designed by the Institute of Energy Conversion at the University of Delaware, gathered 80 percent of its heat and electricity from the sun using active solar collectors and photovoltaic cells. Its heating and power plants added only 12 percent to the total construction cost, and even that was considered too much: Carl Boer, the institute's director, felt that the public would accept solar power only if the added cost of installing the gizmos was less than 10 percent of the house's purchase price. Boer must have gasped when he heard about Lorriman's incremental costs. But he was gung ho on solar as an answer to OAPEC: "Solar energy," he said in 1976, "impinging at high noon on an area 35 kilometres by 35 kilometres, equals the total peak capacity of all existing power plants in the United States combined." In other words, the amount of solar energy falling on the city of Waterloo would be enough to power all of Canada – space heating, hair dryers and all.

Initially, governments and manufacturers were not slow to pour their energies into active solar. By 1977, active had become big business: that year, Solar Power Corporation, a subsidiary of Exxon, sold $15 million worth of solar technology; ITT was manufacturing a solar hot-water heater; General Electric was making solar collectors; Westinghouse was selling heat pumps (devices that took the heat out of one material and invested it in another); and Mobil Corporation was marketing photovoltaic cells not markedly different from the one stumbled upon by Fritts nearly 100 years earlier. Mass acceptance of renewable energy in the form of active solar seemed just around the corner. As early as 1977, the

Science Council of Canada was predicting that "by 1990, the market for solar equipment, installation and maintenance [in Canada] could be $11 million." The next year, EMR allocated $380 million to encourage the development of active-solar technology. In announcing the plan, energy minister Alistair Gillespie said that his objective was "to create an unsubsidized Canadian solar industry, and it must be done within the next five years, or we've lost our chance."

Ironically, shortly before Gillespie's much-publicized push to active solar, active solar received a serious setback in Saskatchewan Conservation House, which opened in January 1978. Built in Regina by the Saskatchewan Research Council (SRC) and the National Research Council (NRC), Saskatchewan Conservation House was Canada's answer to Solar One, designed primarily as a demonstration project to show prospective homeowners and builders that huge energy savings were possible. As its name implies, the thrust of the house was in saving oil, not replacing it. The rectangular, two-storey, 2,000-square-foot house was equipped with mechanically operated window shutters insulated to R-15 to control heat loss during the night from the seven south-facing windows, and the outside walls—built as "double walls," a system invented by NRC's Harold Orr specifically for this house—were 12 inches thick and insulated to R-40, with R-60 in the ceiling, three times the standard for the time.

The walls of Saskatchewan House were wrapped on the inside with 6-mil polyethylene, double-seamed and caulked so that virtually no air could move through them. The house was so tightly constructed, in fact, that nearly half of its heat could have been provided by its south-facing windows. Nonetheless, eight solar panels, which cost $30,000 and were made in the United States, were installed on the house's roof, and a 2,800-gallon water-storage tank was lodged in a room that ran along the west wall (there was no basement). The system was there at the insistence of NRC, which was still promoting active solar.

It broke down almost immediately. In an assessment of the house's performance published in the May 1979 issue of *Solar Age*, Rob Dumont, who worked on the design as a graduate student, reported that "the active-solar system has been troublesome. The thermostatic controller malfunctioned, causing the collectors to run day and night. The temperature sensor in the collector burned out. The glycol solution boiled over due to pump failure, and solenoid valves controlling heat distribution to the house malfunctioned." All in all, the house was a poor ambassador for active-solar technology.

Fortunately, the failure of the active system to heat the house was

19

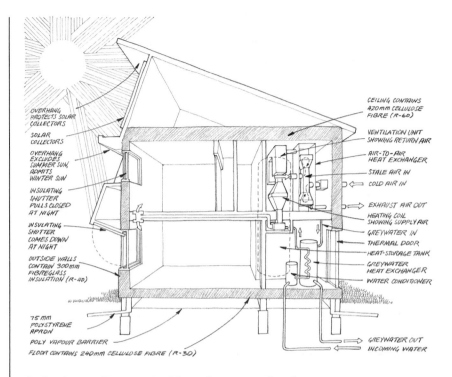

CEILING CONTAINS
420mm CELLULOSE
FIBRE (R-60)

OVERHANG
PROTECTS SOLAR
COLLECTORS

VENTILATION UNIT
SHOWING RETURN AIR

SOLAR
COLLECTORS

AIR-TO-AIR
HEAT EXCHANGER

OVERHANG
EXCLUDES
SUMMER SUN,
ADMITS
WINTER SUN

STALE AIR IN

COLD AIR IN

INSULATING
SHUTTER
PULLS CLOSED
AT NIGHT

EXHAUST AIR OUT

HEATING COIL
SHOWING SUPPLY AIR

INSULATING
SHUTTER
COMES DOWN
AT NIGHT

GREYWATER IN

THERMAL DOOR

OUTSIDE WALLS
CONTAIN 300mm
FIBREGLASS
INSULATION (R-40)

HEAT-STORAGE TANK

GREYWATER
HEAT EXCHANGER

WATER CONDITIONER

75 mm
POLYSTYRENE
APRON

POLY VAPOUR BARRIER

GREYWATER OUT

FLOOR CONTAINS 240mm CELLULOSE FIBRE (R-30)

INCOMING WATER

Saskatchewan Conservation House demonstrated to the housing industry that insulation, airtight construction and multiglazed windows would result in huge savings: after one year, the house's total energy bill was $62.96.

balanced by the success of the passive solar gain. Sunlight entering through the relatively small south-facing windows accounted for 44 percent of the house's heat load; occupation—heat gathered from lights, motors and the operation of appliances—added 15 percent; and a backup electric heating system supplied the rest. Conservation measures and the passive solar gain had reduced the house's heating requirements to an unbelievable 5 million BTUs, compared with the average for conventional prairie houses of 171 million BTUs. The house's entire energy bill for the year, including hot water, appliances and backup heating (but excluding the active-solar system), was $62.96.

Saskatchewan House was a success after all. For one thing, it attracted 37,000 visitors the year it was open to the public, which meant that it wasn't just engineers and bureaucrats who were interested in ways to reduce energy consumption. In its own serendipitous way, Saskatchewan House was a milestone on the long, arduous road to Waterloo.

"Saskatchewan House deserves its position in this whole thing," says Oliver Drerup, a low-energy builder who was mightily impressed by Orr's concept. "They did an amazing thing: they proved that what they were trying to prove was stupid and that there was something else that was more important." What the house proved was that with superinsulation, passive-solar design and $63, you could run a house for an entire Saskatchewan winter. In 1978, that was no small thing.

For those turning their backs on the technofix and treading the soft path, there was more good news to come: the $60,000 solar panels on Lorriman's house were not working well. Designed to contribute 75 to 80 percent of his space-heating needs, they were, in fact, providing barely 50 percent. Square foot for square foot, more heat was coming from a pair of glass sliding doors leading out to the backyard and a small greenhouse off the kitchen, both of which were there simply to provide light to the kitchen. They were, in effect, a passive solar heat source and were contributing 25 percent of the huge house's space-heating requirements, proportionally 10 times more than the active-solar panels were providing, at about one-hundredth the cost.

There were a few people around who had known this all along. Charles Simon, for one. Simon, sometimes called the grandfather of passive solar in Canada, was designing and building sun-heated houses even before OAPEC. "We didn't call it passive solar in those days," says Simon. "It just seemed like common sense to take that usable heat and draw it into the house." In 1973, Simon designed a sun-heated house in Toronto that incorporated two wedge-shaped towers joined by a greenhouse walkway. Although the greenhouse was intended to supplement a basement furnace, it ended up supplying almost all the heat for the entire house. Simon placed heating vents in the greenhouse for the winter and insulating curtains for the summer, but they were never used: the greenhouse was allowed to fluctuate with the seasons, and only when the outside temperature dropped to 20 degrees F did the furnace thermostat kick on. In fact, a ceiling fan in the greenhouse, meant to expel excess heat in the summer, often came on during the winter. This surprised even Simon. "Then," he says, "it was all guesswork."

Greg Allen was another innovative low-energy designer who was working with solar in the early 1970s. The first house he designed, built in Gananoque, Ontario, had a sod roof with a 240-square-foot active solar collector on it. The system worked all right but provided only about half of the house's heating requirements, the rest being made up by the large south-facing windows that poured light and heat into a greenhouse and by a wood-burning cookstove and a Franklin fireplace. Allen, afraid that the active system was too technological for homeowners without a Ph.D., figured he could go a lot further in consumer acceptance with passive solar. "It's unfortunate," he said, "that so many of us have lost a fundamental understanding of the things that keep us warm."

His second house was basically a huge passive solar collector, with the entire southern exposure covered in glass at an angle of 45 degrees. The radiant energy that came in through the glass was collected and stored in the concrete floor slab and in the 22-ton beachstone fireplace that was the building's centrepiece. This system supplied 80 percent of the house's heat, with the fireplace making up the rest. Allen realized that the secret to making passive solar work was controlling the temperature swings between day and night and between sunny and cloudy spells; storing heat in the masonry solved part of the problem – the massive floor warmed up to 95 degrees F on sunny days and stayed warm long after the sun went down at night.

But Allen realized that the best way to control the swings was with insulation. Allen insulated Prospect House with bales of hay lined with 8 inches of Styrofoam or 12 inches of fibreglass batts, for an R-value of 20. The total cost of construction was just $35,000, and the house could be heated with a single cord of fuelwood a year.

□ □ □ □ □ □

IN ORDER TO HAVE A LOT OF GOOD IDEAS, YOU ALSO HAVE
TO HAVE A LOT OF BAD IDEAS.
— THOMAS EDISON

Meanwhile, a similar light was dawning on three builders – one of whom would later become involved in the Waterloo Greenhome – in Quebec's Eastern Townships in the early 1970s. The builders were Elizabeth White, Oliver Drerup and David Kantor. Elizabeth came to Canada from England in 1969 with a degree in architecture and spent her first two winters in Montreal, at McGill University. She encountered snow, which changed a lot of her ideas about architecture. "I had never seen snow like that," she says, "literally piled up over your head. It made me appreciate the impact climate has on your life: it was a shock exposure to microclimate design."

She also encountered Oliver Drerup, a force at least as powerful as snow. Oliver was a teaching assistant in archaeology at the time, but his mind was ranging around in a number of unrelated areas, one of which was the energy crisis. The two of them met David Kantor, a student at McGill who was interested in sensible architecture. "Passive solar hadn't entered our consciousness at all at that point," says Elizabeth. "My own architectural thinking was to restrict the window area to specific views without worrying about the direction." But it quickly became apparent to all three of them that sensible architecture was low-energy architecture.

They eventually met Nick Nicholson, an active-solar enthusiast who had settled in the Eastern Townships at about the same time the United States settled into Vietnam. Nicholson was experimenting in both active and passive—he had built an active-solar-heated doghouse on his property near Ayer's Cliff and had written one of the first textbooks on active solar. Elizabeth, David and Oliver were reading everything they could find about alternative energy in those days, which wasn't a lot, and when they read Nicholson's book, they became hooked on active. So when in 1976, Ken Elliott, a colleague of Oliver's at McGill, asked their fledgling company, DKW Designs, to design and build a house for him, they went to full-blown active solar.

The house was to be located near the village of Stanstead. "It was an interesting site," recalls Elizabeth. "It was north-facing, looking over a cornfield, so we covered the entire south wall with solar panels. We put a rock storage bin under the floor and invented a reversing damper that would dump the heat in at the top of the rock storage and take it out at the top too, instead of at the bottom, so that you could extract heat before the whole store was fully charged." They also ran copper tubing through the storage bin to preheat water before pumping it to the domestic hot-water heater. "It all worked pretty well," she says.

It worked, but it didn't take them long to realize that it could have worked better. "Oh God," says Oliver, "I remember it so well. We had finished the house and were all sitting around in the kitchen congratulating ourselves. We decided to calculate how much insulation we could have purchased with the same amount of money we had spent on the collectors, and then what the heat load for the house would have been with that much insulation. It so overwhelmed the active-solar contribution that the three of us just sat there utterly stunned by the implications. We thought, holy shit, we've just done something really stupid."

It was an epiphany, the kind of thing Paul is supposed to have had on the road to Damascus: the secret of passive solar was insulation. Lorriman's house had received 25 percent of its heat from its passive components, and it had very little insulation in its single-frame walls—4 inches of fibreglass batts, which was still more than twice the amount called for in the National Building Code. Once builders learned how to keep 23 the heat in and the cold out, the advantages of passive became apparent. Except for the odd fan here and there, it was nonmechanical, which meant there were fewer things to break down. It was a lot cheaper than active. The glass and design features were not extras, and the costs of thicker walls and additional insulation were quickly bought back by lowered heating bills. And it was easily accessible to the average builder: various types of insulation were sold in hardware stores, and anyone could install it once the principles of air movement were grasped.

There were, of course, problems with passive solar. Today we call them challenges. The main challenge had been obvious from the start: passive solar heating was fine when the sun shone, not so fine when it didn't. At first, this was thought to be the same problem active solar had – how to store the heat gained during the day and distribute it throughout the house at night – and consequently, many of the early passive solutions were adaptations of active solutions. Charles Simon's second house, for example, combined the mechanical gewgaws from an active system – fans, a heat exchanger, a huge rock storage bin buried under the house – with a simple solar collector (a greenhouse). Instead of combining the best of both worlds, it dramatically demonstrated the failures of each: the greenhouse was insulated from the rest of the house, and the rock storage bin took nearly six months to heat up.

But essentially, the light on the Damascus road had been seen, and it wasn't about to be forgotten. In 1970, the United States Department of Energy reported that there were only 24 houses in the entire nation specifically designed to take advantage of solar energy; by 1978, there were 40,000 such houses. But more important, the passive approach had won: designers felt less and less obliged to scramble about looking for a flashy new source of heat to replace oil; they just had to find a way to hold onto the heat a house naturally absorbed from the sun. And that soft-energy path leads directly to the Greenhome.

□ □ □ □ □ □ □

By changing the house, we ultimately change society.
— Sean Wellesley-Miller, *Toward a Symbiotic Architecture*

The effort of trying to make hybrid systems work served mainly to drive home the simple fact that if you have to jump through too many hoops to solve a problem, then you're probably running at it from the wrong end. The real drawback was quite simple: no matter how much heat entered the house through the windows and no matter how ingeniously you tried to keep it there, if more heat left through the windows than entered through them, the house would be cold. The best windows available at the time were double-glazed units with aluminum spacers to keep the two panes apart. Such windows still represented an overall heat loss to the house, because with an R-value of only about 2.4, they conducted heat out as soon as the outside temperature dropped lower than the inside temperature. Even south-facing windows provided only a minimal net heat gain. According to Rob Dumont, who is now one of the consultants with the Advanced Houses program, "the net solar gain (incoming solar radiation minus heat losses) for south-facing windows in a cold climate can be quite small. For instance, a standard double-glazed window facing south in a Saskatoon home will actually have a net heat loss over the months from November to February. Adding more south glazing actually makes the heating performance of the house worse." In 1979, Dumont proved his point. He monitored the energy performance of two houses similar in all respects except that one had 32.5 square metres of south-facing glass while the other had only 11.7 square metres. The house with the larger windows cost more to heat.

Dumont divides the passive-solar movement in the late 1970s into two distinct groups: those who took the "mass-and-glass" approach and those who opted for "light and tight." The former favoured large, angled, south-facing windows with lots of big, solid things inside to soak up and hold the heat — concrete slab floors, stone fireplaces, metal furniture, ceramic-tile tables, Trombe walls, Plexiglas tubes filled with water (and sometimes fish), ceilings that were actually huge waterbeds. We've all seen them. We've all driven through the countryside on a road that passes through a lovely stand of mixed hardwoods, and suddenly over the treetops rises a sharp, jagged roofline, a spear of redwood and glass, as if thrust up toward the sun by a disturbance deep in the Earth's core. They are often spectacular. They usually cost a lot. Sometimes they even work.

Light-and-tight proponents followed Harold Orr in opting for smaller, less expensive houses, with modest south-facing windows that allowed in enough sunlight to provide space heating and, to keep the heat in, double-frame walls, a well-sealed envelope and lots of insulation. Lots and lots of insulation. Twelve inches of insulation in the walls was considered barely adequate. Saskatchewan House was light and tight and had 12-inch walls. So did a house built about the same time in Massachusetts by Eugene Leger. The Leger house had double walls and so much insulation that it was heated by the excess heat given off by its hot-water heater. It didn't look very spectacular, though. It looked like a large shoebox with a peaked roof. But it sure was cheap to run.

Builders liked building light-and-tight houses because they were easy — they didn't have a lot of weird, knife-edge angles jutting up into the sky that required complicated cuts (and therefore extra labour) and lots of wastage — and because they looked so much like conventional houses that they were a snap to sell even to buyers who didn't give a hoot about the environment. Theorists liked light-and-tight houses because, well, because they were light and tight. They worked. Harvard physicist William Shurcliff reviewed Leger's shoebox design and hailed it as an entirely new kind of system, a concept "just being born." He liked its "truly superb insulation. Not just thick, but clever and thorough." It was effective not only in the easy spots but also around "sills, headers, foundation walls, windows, electrical outlet boxes, et cetera." He praised its absence of extra-large thermal mass (a body blow to the mass-and-glass defenders): "Down with Trombe walls!" he cried. "Down with water-filled drums and thick concrete floors!" He also liked Leger's house because it was "no weird shape of house, no weird architecture," which was true. "What name should we give this new system?" he asked rhetorically. "Superinsulated passive? Supersave passive? Mini-need passive? Micro-load passive? I lean toward 'micro-load passive.' " Fortunately, the rest of the low-energy community leaned toward "superinsulated."

As the superinsulated people say, insulation is a material, superinsulation is a system. Loading a house with astronomical R-values won't keep the heat in if warm air can convect to the outside through massive gaps in the blanket. Such gaps are commonplace in conventionally constructed houses. Fibreglass batts stuffed between 2-by-6 studs are not always tight to the studs and can slouch down over time to leave empty spaces at the top plate. Superinsulated houses incorporate improved installation techniques, so that the gaps do not occur, as well as an interior wrapping of polyethylene so well sealed that virtually no air at all passes through the walls or around the doors and windows. The term is "airtight." In a conventional house, if all the incidental gaps in the building shell were joined together, they would constitute a 144-square-inch hole in the wall — like having a 1-by-1-foot window open all winter; in the Waterloo Greenhome, that hole would be about the size of a looney.

The mass-and-glass people didn't give up, of course. A mass-and-glass house, too, could be superinsulated and airtight and still have thermal mass and lots of glass. But even they had to admit that in terms of thermal leakage, the weakest part of their building system was not around the windows but through them. This weakness was accentuated in a superinsulated mass-and-glass house: having three R-40 walls was all right, but if the entire south-facing side was glass, which has an R-value of less than 1 per single pane, then the house was still going to invite comparison with a flour sifter.

So the question among mass-and-glassists became: what do we do about the windows? The obvious answer was to find some way to insulate them at night, to prevent the heat gained during the day from returning to the great outdoors when the sun went down. Early window covers ranged from the sublime to the ridiculous, as various kinds of cumbersome appendages were added to the windows of a great number of otherwise beautiful houses. Imagine a soaring, lofty, cathedral-ceilinged living room, softly lit, fire in the airtight stove, classical music on the stereo, lovingly crafted furniture arranged around carefully chosen antiques, and six hulking great foil-lined futons strapped against one whole wall and rolled up every morning to hang over the room like a threat of bankruptcy. That was one solution. Some people built wooden frames enclosing squares of foil and rigid foam insulation that would fit into the window cavities at night. But what to do with them during the day? Well, hoist them up to the ceiling on pulleys, or better yet, cover them with some nice tie-dyed quilted material and hang them on the other three walls so that your living room looked like a padded cell for Andy Warhol. Other ingenious solutions appeared on the market: double-glazed windows fitted with electronic gizmos that slid thin sheets of aluminized fabric into the space between the panes or blew in millions of tiny beads of pulverized polystyrene and then sucked them out again in the morning.

The problem with most of these methods, apart from aesthetics, was that they required an inordinate amount of owner participation. Somebody had to be home at night to crank down the padded covers or blow in the polystyrene pellets or take Andy Warhol's nightmare down from one wall and jam it into the window frame. And somebody had to get up in the morning to undo the whole thing, or else as soon as the sun hit the insulated, double-glazed, very expensive windows, the heat differential between the inside and outside panes would crack them. How many people wanted a house that forced them to be home by sundown and up at sunrise every day? People began to feel chained to their houses. It was worse than having a dog.

By and large, window covers didn't work all that well at the best of times. No window cover could be airtight or deliver as high an R-value as a wall. The best Saskatchewan House could do was install quadruple-glazed windows (which the builder made on-site by caulking two double-glazed units together) on the north wall for a value of R-15. Quadruple panes of glass on the south-facing windows would have blocked enough solar gain to defeat their whole purpose. In 1979, the Canada Mortgage and Housing Corporation (CMHC) devised an interim set of standards for movable window insulation and made approved types eligible for grants under the Canadian Home Insulation Program (remember CHIP?). But as John Haysom, an energy consultant who had helped define CMHC's criteria, admitted, the standards "involved a certain amount of guesswork, so they are to some extent arbitrary." Even so, by 1983, only one product—something called Window Quilt, made in the United States—had passed the not-very-rigorous CMHC exam.

Well, a building had to have windows; it said so right in the Code. So if you couldn't cover them up and you couldn't get rid of them, then the only thing to do was make them better. The first step was to figure out what it was that made conventional windows inefficient. Glass is an extremely poor insulator: a single pane of glass has an R-value of 1, and even that is due largely to the microscopically thin film of air that adheres to its two surfaces. The main advantage of glass as a window material (apart from the engineeringly insignificant fact that you can see through it) is that it transmits visible (short-wave) light but absorbs or reflects thermal (long-wave, or infrared) radiation, which is why you can see the sun through a window but you can't get a tan through it. And short-wave light can pass through glass by transmission, but once that energy is turned into heat, it can escape through the glass only by conduction.

The trick to making windows work, then, was to somehow inhibit the movement of heat from inside the house to the outside without sacrificing the transmission of light from outside to the inside. Since the best-known inhibitor available at the time was air, the first window improvements simply increased insulation values by increasing the amount of air around the window. That was the theory behind double-glazed windows: the airspace between the panes increased—in fact constituted—the window's insulating properties. Second-generation improved windows were simply triple and quadruple glazings. Things bogged down here for a while, because, as noted above, the amount of glass involved in quadruple glazing inhibited solar transmission more than it retarded thermal conduction. The windows themselves also became somewhat unwieldy. Then things like special coatings for the glass and special gas fillings for the airspace between them came along, and the tempo picked up again.

This is pretty much where low-energy house design was by the end of the 1970s: innovators had encountered many of the problems, and engineers were busy trying to solve them. Although more and more houses were being built with environmental impact high up on the priority list — as it is in the Waterloo Greenhome — there was still a large gap between the availability of such information and the number of times it was being put to use.

One of the reasons for this — and it's a problem the Greenhome designers had to grapple with — was cosmetic: although Saskatchewan House had been a brilliant technical success, many of those who visited it said they didn't like the way it looked. The interior was laid out oddly, they told the people conducting the tours: it didn't have a basement, the mechanicals room was on the second floor, and from the outside, it looked, well, boring. It didn't look as boring as the Leger House, but it didn't look much more exciting than any other new house on the block either. The question of cosmetics goes right to the heart of public acceptance.

In response, Saskatchewan's Office of Energy Conservation launched a large-scale building program called the Energy Showcase Project, a competition set up in 1980 and overseen by the same team that had designed Saskatchewan Conservation House: Harold Orr, Robert Besant and David Eyre. The idea was to encourage the design and construction of low-energy houses that would appeal to the general buying public — to prove, in other words, that houses didn't have to sacrifice looks to have brains. Fourteen houses by 14 different designers and builders were given the green light in Saskatoon.

One of the main innovative features of the project was that rather than specify the kind of heat source the houses could use, the program simply set a series of performance standards. In other words, the Office of Energy Conservation said to the builders, "Build us a house that uses about one-third of the energy consumed by a conventional house — that is, no more than 55.5 kWh per square metre of floor space — and we don't care how you do it." Some of the innovations the builders came up with were primitive but effective. One house, taking a page from Leger's book, used a system in which a high-efficiency hot-water heater also heated the house; another borrowed Saskatchewan House's idea of quadruple-glazed windows; a third was superinsulated to R-60; another used a smaller variant of the Trombe wall. But all of them proved that low-energy features were compatible with consumer-acceptable design.

The other long-term benefit from the Saskatchewan Showcase was a computer program devised by the technical team that could predict how much energy a house would require. Feed in such design features as wall thickness, type of insulation, ratio of window area to floor space, et cetera, and the computer would spit out a number; if that number was less than 55.5 kWh/m², then the house passed the test. The program, called HSLOAD, was so successful that computer hackers across the country came up with similar software almost overnight. When NRC funded a survey of the field in 1983, they found more than 400 performance-predicting programs on the market.

The Saskatchewan Showcase was a major step toward mass-market acceptance of low-energy housing, but it didn't go far enough. In fact, it didn't go anywhere at all outside Saskatoon. The 1970s had produced some giant strides in the labs and basements of the nation, but the accumulated information was still far from becoming general knowledge. Somehow, the word had to be spread to a larger audience. What was needed was a national program.

The same year that Saskatchewan announced the Showcase Project, that is 1980, EMR introduced a similar but bigger and more ambitious initiative: the R-2000 program. Launched in conjunction with the Canadian Home Builders' Association (CHBA), the R-2000 program was intended to encourage every homebuilder in the country to build low-energy houses. Like the Saskatchewan Showcase, R-2000's emphasis was on overall performance: an R-2000 house had to have an annual space-heating budget of under 60 kWh/m², and EMR didn't care how that was achieved. It promoted high insulation values (walls had to be at least R-20), airtight vapour barriers and mechanical ventilation systems that warmed incoming cold air with outgoing warm air. This was light and tight with a vengeance. Any new house that passed the R-2000 test earned its builder a $6,500 rebate from EMR to offset the extra time and materials required to meet the standards (installing a proper air/vapour barrier, for example, can add several days to a contractor's payroll).

R-2000 was basically an education program for homebuilders. Its aim was to make it economically feasible to build light, tight, low-energy houses, houses that would be to conventional building what the Volkswagen beetle had been to Detroit. EMR set up training seminars led by converts like Oliver Drerup to show builders how to make a house airtight enough to pass the R-2000 tightness test of 1.5 air changes an hour at 50 pascals of air pressure (or about 0.5 air changes per hour under normal conditions). Since the program's inception, more than 15,000 builders have been trained in R-2000 methods, and it has been a model for countries around the world. According to Oliver, the goal was to turn the construction industry's collective mind from short-term thinking to long-term planning. Too many people in the industry, he says, were just not bothering with low-energy construction because, to use the buzzword of the '80s, it wasn't cost-effective. "Cost effectiveness is a very difficult issue," he says. "What's the cost effectiveness of a cruise missile?"

The holy trinity of the R-2000 program was superinsulation, airtightness and ventilation; its credo was conservation. Perhaps Canada was following the example of the energy movement in the United States, where, in the words of Arthur Rosenfeld, who set up the American Council for an Energy Efficient Economy in 1979, "improved efficiency" of energy use in the States "is saving us $140 billion per year in energy costs," a sum that amounted to $1,750 per U.S. household. Canadians wanted in on those savings, and the way to get in seemed to be through conserving energy rather than creating alternative sources of it. In their 1981 book, *Energy and the Quality of Life: Understanding Energy Policy*, a group of University of Toronto economists echoed virtually every other contemporary study when they noted that in Canada, "the potential for energy conservation is so great that with only modest life-style changes yet with modest economic growth, medium-term energy-demand levels could be held to present levels or perhaps even reduced by applying the appropriate conservation measures." Modest life-style changes; held to present levels. Don't panic.

"Without conservation measures," they warned, "no practicable policy of increased supply—renewable or conventional—can realistically and acceptably meet future energy demands." And therefore, "investment in energy-conservation measures is in general significantly more cost-effective"—there's that phrase again—"than any alternative investment in increased energy supply." If we have money to dump, in other words, better to dump it into conservation than into developing a renewable-energy industry in North America.

Hence the conservation focus in the R-2000 program and in all subsequent low-energy initiatives, including the Advanced Houses program. Solar energy and natural gas are still components, but the emphasis is definitely on reduction rather than revolution. Still, R-2000's potential to fire the imagination of the ordinary homebuilder was amply demonstrated by a tract-house builder in Winnipeg named Rubin Diamond. Diamond visited each of the 14 low-energy demonstration houses in Saskatoon and came away feeling that he had been granted a vision of the future, and the future was airtight. Back in Winnipeg, he invited Harold Orr to talk to his architects. Orr's philosophy was simple to understand and hard to refute: build double-frame walls capable of holding lots and lots of insulation, line them with a 6-mil polyethylene vapour barrier that was as continuous and unpunctured as possible, and the house will consume 50 percent less heat energy than a house built to

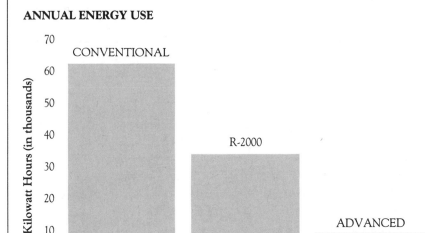

ANNUAL ENERGY USE

Kilowatt Hours (in thousands)

CONVENTIONAL

R-2000

ADVANCED

ENERGY PERFORMANCE

Code standards. "Half of a house's heat loss," Orr told Diamond's team, "is through air leakage and air infiltration. Reduce those two things, and it stands to reason that you reduce your heating budget."

Diamond took Orr's words to heart and decided to make the Flair subdivision outside Winnipeg Canada's first entirely energy-efficient suburb. The first superinsulated Flair house opened in January 1981. It had R-44 walls and R-60 ceilings. It had an air/vapour barrier so tight that a quarter-inch hole in the electrical wiring intake box roared like a hurricane in Diamond's ears and he had it plugged. His workers used three dozen tubes of Tremco caulking on each house. They buried the barrier between the double walls so that drywallers wouldn't puncture it with their screws. They sealed the ceiling before installing the partition walls so that the ceiling barrier would be a single unit. The result was almost unbelievably tight. In a conventional house, the inside air can be replaced up to 10 times an hour by cooler outside air coming in through leaks in the building shell; in a typical R-2000 house, air changes are reduced to 0.5 an hour. In Rube's Baby, as the first Flair home came to be called, the air change was 0.12 an hour. Translated into practical terms, if the house next door to Rube's Baby cost $600 a year to heat, the Babe came in at around $100. To be on the safe side of incredible, Diamond advertised his Baby as "the $150 house."

Unfortunately, the R-2000 program didn't encourage many builders to emulate Rubin Diamond. In fact, according to Elizabeth White, keeners like Diamond were actually *discouraged* from trying to beat R-2000 standards. There were no increased incentives—no larger rebates, no Awards Banquet. "There were many, many small builders out there who were truly enthusiastic about R-2000," says Elizabeth, "who wanted to beat that air-change standard, but the trainers were telling them that if their house was more than meeting the standard, they should reduce the insulation. It wasn't cost-effective. And over time, it resulted in the keeners being less enthusiastic." The idea was that if the standards were made to seem easy to meet, then the larger builders would jump on board. "I don't think that's happened," says Elizabeth, and the numbers bear her out: in 1991, 11 years after the introduction of the program, only 4,000 R-2000 houses had been built in Canada—fewer than 400 a year—during a period when annual new-house starts averaged 125,000.

Why? Ironically, part of the reason may be the easy-to-meet standards: people like challenges, and once the principles of airtightness were grasped, anyone who could squeeze a caulking gun could build an R-2000 house. A lot of houses being built today incorporate R-2000 practices without being registered officially as R-2000 houses, and in 1995, many R-2000 standards will be written into the National Building Code. But in general, it's fair to say that interest in the program peaked early.

In the low-energy community, however, the flame was kept alive, although it was flickering ominously. If conservation was to be our god, they decided, then let it be conservation with a capital C. In 1983, NRC's Division of Energy approached Greg Allen, Oliver Drerup and Elizabeth White—who had teamed up and were designing and building extremely interesting low-energy houses in the Toronto area under the name ADW—with a singularly intriguing assignment. The modern term for it would be "blue-skying." They were asked to design the most energy-efficient house imaginable. They were to pretend that there was an unlimited budget, that they were building on NRC property with no lot specifications to worry about and that technology existed which at the moment was only an engineer's pipe dream. Think prototype, they were told. Think International Energy Agency (IEA), an august group of low-energy designers from around the world, a sort of energy-efficient United Nations, of which Canada, via NRC, was an active member. What would the purchased energy consumption in such a house be? How close to zero could they get? "Real-world conditions weren't valid," says Tim Mayo, who was indirectly involved as an NRC researcher at the time. "Just push the technology as far as it would go."

What ADW did in response was invent an advanced integrated mechanical system (IMS), a single unit that took care of the heating, cooling, hot water and ventilation all in one compact piece of hardware, and contain it in an airtight, superinsulated, passive-solar envelope. They included a lot of commonsense items in the house as well—fluorescent lights, low-energy appliances—whose main thrust was cutting energy use to the bone. Their design was the base on which the Waterloo Greenhome was built: in fact, as Tim Mayo says, "What Saskatchewan House was to the R-2000 program, ADW's design was to the Advanced Houses program." When they ran the design through NRC's HOTCAN computer program to find out what its energy budget would be, the prediction astonished even them. "The average new house at that time," says Tim, "used the equivalent of something like 60,000 kWh to 80,000 kWh of electricity a year. The same house in an R-2000 version would come in at 35,000 to 40,000. This house came in at 4,500 kWh a year. One-tenth of the most energy-efficient house then on the market! One-eighteenth of a conventional house!"

When Tim flew to Stockholm that August to present the ADW data to a meeting of IEA designers, "they didn't believe me," he says. "I told them, 'Here are the energy figures: 4,500 kWh per year.' And they said, 'Okay, but how much for appliances? How much for lights?' And I said, 'No, no, they're all in here—this is total.' And their jaws dropped."

It was a fine moment for Tim, a fine moment for ADW, a fine moment for Canada and a fine moment for the environment. A few months later, NRC announced it was closing down its Division of Energy, citing as its reason the withdrawal of funding by the Conservative government. Once again, the long road from OAPEC to the Waterloo Greenhome had taken an abrupt turn.

GROUND WORK

□

WE SHALL NOT CEASE FROM EXPLORATION
AND THE END OF ALL OUR EXPLORING
WILL BE TO ARRIVE WHERE WE STARTED
AND KNOW THE PLACE FOR THE FIRST TIME.
—T.S. ELIOT, *LITTLE GIDDING*

WHAT'S GOING ON? TODAY'S DREAM HOUSE IS LOOKING
MORE AND MORE LIKE THE HOUSE OF A HUNDRED YEARS
AGO.
—WITOLD RYBCZYNSKI, *LOOKING AROUND: A JOURNEY
THROUGH ARCHITECTURE*, 1992

From Tim Mayo's corner office on the seventh floor of the EMR building in Ottawa, you can, if you crane your neck far enough to the left,
see the old Dominion Experimental Farm, a huge tract of farmland in
the middle of Ottawa. I've always thought of it as a kind of symbol of
Canada, this farm at the heart of the nation's capital. It is, however, as
much living laboratory as working farm. Generations of agricultural
scientists have tracked mud through its maze of barns, greenhouses and
office buildings, developing new crops, solving farmers' problems and
generally rummaging around in the inner workings of nature. In 1910,
Charles Saunders, later Sir Charles Saunders, crossed 'Red Fife' and
'Hard Red Calcutta' wheat in one of those greenhouses and produced
a cold-hardy, disease-resistant, high-yield strain of wheat he called 'Mar-
quis,' which for the first time allowed wheat to be planted in Saskatchewan and Alberta, making Canada a world leader in wheat production
and export. This is something that Tim, when he came back from the
IEA meeting in 1984, thought was about to happen in low-energy technology, and in a roundabout way, it has. Canada is a world leader in
cold-weather technology, especially as it applies to shelter; for Canadian
low-energy designers, however, that reputation hasn't been easily gained.

Tim started working for the NRC as a consultant when the grass-roots
low-energy movement was in full swing. People like Greg Allen and
Elizabeth White and Charles Simon and hundreds of others were building houses and inventing weird gizmos that worked and, with a little
government funding, could have gone into full production. It was happening in the United States and in Sweden. It started to happen here.
In 1982, the Trudeau government instituted such popular conservation
programs as the Canadian Oil Substitution Program (COSP) and the

36

Canadian Home Insulation Program (CHIP), which helped homeowners switch from oil to other forms of heating systems and to upgrade the R-values in their houses. It established Canartech Conservation Inc., a division of Petro-Canada that provided venture capital to people wanting to invest in conservation and renewable-energy products. And it funded some 30 projects in low-energy research in its various federal departments, all directed by a central secretariat called the Panel on Energy Research and Development, or PERD. Things were looking rosy and continued looking that way until the election campaign of 1984.

During that campaign, the Conservative Party promised to double federal research spending on renewable energy. When Finance Minister Michael Wilson delivered his first budget speech that November, however, he announced a budget cut of $4.2 billion, almost half of it to come from environmental programs, which included research into renewable energy. COSP was cut; CHIP was reduced; Canartech was scratched. The NRC's Division of Energy, which had been filtering $30 million into solar research, was also cut.

Wilson's paper was entitled "A New Direction for Canada," and from a low-energy point of view, that new direction was clear: along the hard-energy path. Consider this: in its first year of governing, the Conservative Party cancelled Canartech, which was funding the search for alternative fuels; raised the price of gasoline by 2.5 cents a litre; and bought Gulf Oil, the largest oil and gas company in Canada.

There was a mad scramble to salvage some of the projects that the Division of Energy had originated. Tim went to EMR, where he was put in charge of passive solar research and development, with a budget of $1 million. By that time, however, so many programs had been cut that few were applying for funding. "When I was hired, that million dollars had two $10,000 contracts on it," Tim says, "and the system in Ottawa is, if you don't spend it, it disappears back into general revenues."

So Tim spent it. He established a kind of unofficial national energy policy – soft, of course – and administered it himself. "I knew about the integrated mechanical system that Greg Allen had designed for the ADW house, for example," he says, "and I salvaged it." With Tim's money, Allen continued to improve his mechanical wonderchild, which he called Solmate. When it was more or less ready for a field trial, ADW went knocking on doors again. "They had a design for this world-class, world-leading house," says Tim, "and no money to build it with." They convinced Ontario Hydro to finance part of the construction, the Canadian Home Builders' Association also kicked in, and Tim persuaded EMR to finance the rest: "I was already funding the mechanical system," he says, "and the goal of the mechanical system was to put it in a house, so I had to fund the house."

The EMR and Ontario Hydro, however, put conditions on their involvement. "The original plan had been to build a house on NRC property that would push technology to the limits," says Tim. "That was dropped. The new proviso was that it had to involve the construction industry. They said, 'You have to be plugged into the mainstream industry; that's who we're trying to reach.' Well, the Home Builders' Association looked at ADW's original design and said, 'What is this shit? We can't support this design. A, it's ugly. B, it has no basement. C, it's got the mechanicals room on the third floor. D, it needs a lot that's 80 feet wide! This is not a real-world project.' So ADW had to go back to square one; they had to redesign the whole house." Rather than representing the forefront of low-energy design on government property — like a field full of a new type of wheat on the Dominion Experimental Farm — the house suddenly had to look like a typical upscale, market-driven house in a typical modern suburb. Its energy target was set at 20,000 kWh a year: half that of an R-2000 house but more than four times higher than the target its designers already knew it could achieve. The site chosen for it was a regular-sized lot in a suburb of Brampton, Ontario, the hometown of Ontario's then premier, Bill Davis.

In other words, what had happened with the Saskatchewan House in 1977 was happening again 10 years later: when the government took over the funding, it also took over the design. The government's goal has always been to get the low-energy message out into the marketplace by converting the people who build houses rather than the people who buy them. That's why the R-2000 program was aimed at builders. It stands to reason that if 10 percent of the houses built conform to R-2000 standards, then 10 percent of the houses sold will be R-2000 houses. It is a much more difficult proposition to educate consumers to actually *ask* for R-2000 houses. With Brampton House, which was going beyond R-2000 standards, that task would be even more difficult; it would be much simpler to use the house as a showcase to demonstrate to the industry that a house that halved R-2000 energy-efficiency standards could be built to look exactly like any other house on the block — a two-storey, basemented house on a 40-foot lot, a clone of its neighbours on the outside at least.

When it was built, Brampton House was disappointing. The builders continued to make changes on-site that streamlined the building process but undercut the energy efficiency. Almost all the energy supplied to the house was electrical, even though by the time the house opened, electricity was no longer a viable low-energy option. And it was huge—3,500 square feet—at a time when the trend among low-energy designers was toward smaller houses. In other words, the house was built to satisfy the needs of its builders rather than those of its eventual owners. The best that could be said of Brampton House was that it was all right as far as it went; it just didn't go very far.

□ □

THERE IS AN URGENT NEED TO RELEARN THE WAY TO BUILD BUILDINGS THAT FUNCTION SIMPLY AND WITHOUT A COMPLETE SURRENDER TO ENERGY-DEPENDENT SYSTEMS. WE MUST RESPECT THE INDIVISIBLE NATURE OF THE EARTH'S SUPPORT SYSTEMS SO THAT WE CAN SATISFY PEOPLE'S NEEDS WITHOUT DESTROYING THEIR EARTH.
—RICHARD G. STEIN, *ARCHITECTURE AND ENERGY*, 1977

IF IT WERE LEFT UP TO US—ARCHITECTS, BUILDERS, THE ENTIRE CONSTRUCTION INDUSTRY—TO SET THINGS RIGHT, THERE WOULD BE NO HOPE AT ALL.
—MALCOLM WELLS, *GENTLE ARCHITECTURE*, 1981

"The low-energy industry is a small but close-knit group," says Steve Carpenter, bending over a Whistle Dog in Sparky's, an accelerated-food restaurant a few blocks from Steve's office. "I'm always being asked how I get along with my competitors, to which I reply, 'If I wasn't friends with my competitors, I wouldn't have any friends.' " His eyebrows meet in the middle and rise in an amused arch.

Steve usually wears a loose-fitting black suit with a bright red tie to work. A gangly, hipless six-foot-four, he looks like the kind of guy whose shirttail is always trying to leap out of his pants. His black hair is straight and unruly, his beard grows close to his lips, and his eyes burn out from the depths of their sockets, leaving an impression of a man in constant rebellion against his own clothes. He exudes energy. He has very little small talk. If you try chatting over lunch with him about baseball, he'll listen politely and say yes a lot, but his eyes will be glowering out from their caves, and his long, slender fingers will be fiddling with a pen.

He studied mechanical engineering at the University of Waterloo in the 1970s. The university has a co-op system in which four-month study terms alternate with four months of work in the field, and Steve's co-op placement was with Atomic Energy of Canada Limited (AECL). He worked in the design/purchase section, "not designing nuclear containment systems, needless to say," he says, "just ordering nuts and bolts for CANDU reactors."

One of the engineers he worked with was interested in wind-generated power, an unusual interest for an AECL employee at the time. It was 1976, when half the world thought nuclear energy was the only reasonable response to OPEC and the other half were building solar collectors out of tinfoil and copper tubing in their backyards. Steve seems to have landed in a kind of science fiction blend of both worlds.

"This guy bought every book on wind energy he could find," Steve says. "He was going to put a wind generator on his house." The engineer and Steve talked about wind during their lunch hours and coffee breaks, and before long, Steve was learning about wind energy just to be able to keep up his end of the conversation. As a co-op student, he also had to write a technical report at the end of each of his field placements, and after that first year, he wrote his report on the generation of electricity by wind. "And of course," he says, looking down at his pen, "you get caught up in the enthusiasm, and so on."

"And so on" means that he ended up doing his master's thesis on active solar energy, combining his interest in alternatives with his expertise in computer programming to devise a performance-prediction method for active solar collectors. After graduating, he went to work for a Toronto firm called Watershed, which manufactured solar collectors and ran research contracts for the NRC's Division of Energy, which was still going strong. A year later, he decided to strike out on his own, and he returned to Waterloo to start a research and development company, Enermodal Engineering Ltd.

"Our first contract was with the NRC," he says, "to do some performance modelling on passive solar heating for water-storage tanks, those big ones used for fighting fires. It was at the Ford Motor plant in Oakville — the government requires those big facilities to have water storage on-site for fighting fires so that they won't have to tap into municipal water lines. The problem was, Ford had to keep them from freezing during the winter, and they wanted to do it without using electric heaters. Basically, the tank was a huge seasonal thermal-storage system, and the question was, If we took the insulation off it and stuck some glazing on the tank and painted it black, how big would the tank have to be in order for the water to get hot enough during the day that it didn't freeze overnight? The complicating factor was that the tank was cylindrical," he says, grabbing a napkin and drawing a cylinder on it, "so as you put on more and more collectors, they go more and more off-south, until you get one that faces due north and isn't doing much for you. So," he says, his pen really flying now, "we had to figure out what the optimum angle would be, and then . . ."

At the time, the alternative-energy field was still dominated by a lot of earnest but untrained backyard inventors, people putting prototypes together in their basements without much formal grasp of the engineering principles that would make them work efficiently. With his engineering background in solar systems and glass, his connections at NRC and his later contact at EMR with Tim Mayo, whose primary interest was in window technology, it was inevitable that Steve would move Enermodal into the area of high-tech windows. His first big break came when he developed a computer program that would give an energy-performance prediction for a building design: feed in the design data, number of windows, kind and amount of insulation, floor space, framing materials, et cetera, and Steve's program would tell you how much energy that building would require on an hour-to-hour basis. He called the program Enerpass. "R-2000 was using a modelling program called HOT-2000, which did the same thing as ours but could give only month-to-month calculations. Ours did a lot more. It took a little bit longer, but you could factor in and get out a lot more data."

The idea came to him to design a computer program specifically for windows. By this time, the NRC's Division of Energy was history, and another of its former employees, Michael Glover, had started a window-research company named Edgetech. Edgetech's edge was in making windows with better spacers; most people were making them with aluminum spacers, but Glover felt that fibreglass would reduce thermal bridging. "But he needed a performance tool that would prove it," Steve says, "and he asked me to design one. So I did. I called it FRAME."

Like Enerpass, FRAME worked so well that other window manufacturers began asking Enermodal to test their windows so that they could include its performance ratings in their brochures. Then Ontario Hydro instituted a low-energy incentive program that awarded bonus points to consumers installing windows with a low FRAME rating; the Canadian Standards Association began insisting that all new windows have labels on them telling customers what the FRAME rating was; and all hell broke loose at Enermodal. "FRAME suddenly switched from being a design tool to a mandatory rating tool, and suddenly everybody needed it. It was a big jump," he adds. "As a matter of fact, I just checked this morning: it's being used in 18 countries." It is being used so much in the United States that Enermodal has recently opened a second office in California just to handle the demand.

Enerpass and FRAME made it possible for even small builders to meet R-2000 targets, at least on paper. All they had to do was feed their design and material information into a computer, and Enerpass would tell them what the total energy package of the house would be: if it came in too high, they could change a few numbers and run it through the program again. Designers didn't even have to see a piece of insulation or a framing member or window. "It's all done on computer," says Steve. "All the window manufacturers were used to sending their windows out to get leakage or structural testing done on them, and when they heard about FRAME, they'd phone me up and say they wanted to get their windows on the Ontario Hydro list, how do they do it? And we'd say, 'Well, all you have to do is send me your drawings.' And they'd say, 'Yeah, but where do we send the window?' And we'd say, 'We don't actually need the window, just send the drawings,' and there'd be this dead air at the other end of the line. We had a joke going at the office: we'd say to them, 'All right, send us a window that's 22 inches by 43 inches,' which is just the size of the window in my kitchen."

Steve is sensitive to the charge from some quarters that the R-2000 program has been less successful than it might have been because its targets were too easy to meet, partly because of computer modelling. "R-2000 was basically a superinsulation program," he says, picking up a french fry. "You know—stuff the walls and the ceiling full of insulation, and you'd use a lot less energy for heating. In that regard, it was very successful. It showed people that yes, you can build houses that heat for very little money. Where it fell down—that's not quite the word—where it didn't go far enough was in looking at the other energy uses of the house, which is precisely where computer modelling could help it. For example, when EMR went back and monitored some R-2000 houses, they found that the space-heating bill was way down, but the bills for appliances and hot water and cooling and so on were equal to or even greater than those in an average house."

The root of the problem, says Steve, is that most of us still haven't changed the way we think about a house; we haven't considered a house as a complete system, a system that begins when the first tree is cut down to make a 2-by-4 and ends, not when the mortgage is signed but only when the house literally falls down. Such a change could manifest itself in the choice of a house that is big enough to fit our needs and not a cubic foot bigger. The R-2000 program didn't even go that far. "R-2000 houses in general," Steve says, "are custom houses, which usually means houses that are in the city, with a higher square footage than an average house, because if you can afford to build an R-2000 house, you can afford to build a big house. People didn't want it to look as if they were suffering to live in an R-2000 house."

The program, in Steve's view, ought to have addressed the total energy budget of a house—the amount of energy needed to run the appliances, the lighting, the hot-water heater—instead of confining itself to the space-heating requirements. "At the time," he says, "there was an opinion evolving that you could only go as far as R-2000 and to go beyond that was absolutely ridiculous." This was partly because it was thought that to go beyond R-2000's minimum requirements was not cost-effective. "Builders felt that the cost of beating the R-2000 target couldn't be recouped in the selling price of the house." In other words, consumers might pay $2,000 more for a house which tested in at 1.5 air changes an hour, but they wouldn't pay $4,000 more if the house tested in at 1 air change an hour, even though it cost the builder that much to make the house that airtight. The solution, says Steve, was to convince consumers that the long-term operating costs of a house were more important than the increased purchase price.

Brampton House performed well enough, despite the drawbacks quickly pointed out by its critics. It met its energy-use goals, it was comfortable to live in, and it didn't look weird: it had a basement, and it didn't have ungainly solar collectors. It was equipped with energy-efficient appliances, there wasn't a single incandescent light bulb in the house, and there was an integrated mechanical system that was providing not only space heating but also hot water. By any reasonable standard, it proved that low-energy technology could be combined with an attractive, livable, even affordable design to produce a house that an enterprising builder would have no trouble selling. But it didn't appeal to the public, and therefore it didn't change builders' minds about what kind of houses they had to build if they wanted to make a living at it.

The problem was simple: not enough people saw the house. In the entire year that it was open to the public as a demonstration house, fewer than 30,000 people went through it. Compare that with Witold Rybczynski's Grow Home, a revolutionary small house (1,000 square feet) built by Rybczynski and his architecture class on the campus of McGill University in Montreal in 1990: it was open to the public for only three weeks, and more than 10,000 people came to take a look at it. "Brampton House didn't get nearly the publicity we would have liked," says Steve. "Maybe a lot of people from Brampton went through it, but hardly anyone in the rest of the country even heard about the project." And, he might have added, hardly any builders anywhere are using techniques or materials demonstrated in Brampton House.

43

EMR had invested a lot of money in the project, and it was difficult to pinpoint any tangible returns other than the international interest shown in the original design. What was needed was national interest in low-energy housing. Tim's solution was to sponsor a house like Brampton House in every province of the country, to involve more builders and to make the technology available for viewing by more prospective homeowners. Of course, he couldn't just commission 10 Brampton House clones—he had to encourage builders to go beyond Brampton House in terms of energy use and efficiency. The Advanced Houses program was born.

□ □ □

ONE MUST TOUCH-THIS-EARTH-LIGHTLY.
—ABORIGINAL SAYING, QUOTED BY AUSTRALIAN
ARCHITECT GLENN MURCUTT

When I arrived at Lot 92 the morning after the survey, a backhoe was busy excavating the hole for the foundation, and Werner Reiter, the Greenhome's project manager, was setting up his level on the driveway of the Dream Home next door. Werner had already solved one problem. The day before, the surveyors had left written instructions on the side of their wooden stakes—"Cut 7' 0" to bottom of footing"—which meant the excavation for the basement should go down 7 feet from the top of the stake. Werner had checked his drawings and realized that 7 feet was actually the distance to the *top* of the footings, not the bottom, and that the hole should therefore be 7 feet 8 inches deep. "It's the kind of thing that happens," he said, feeling happy he'd caught the error rather than annoyed that it had occurred.

I helped him set up the level by holding a 4-metre calibrated stick atop the surveyors' stake while Werner squinted at it through the level. When he had the stick in his cross hairs, I moved the point of a ballpoint pen up and down on it until he said, "There," and then I marked the spot. Werner took the stick from me and measured 7 feet 8 inches up from my pen mark, then tied a strip of red plastic surveyor's tape at the mark. "We could have used any old stick for this," he said, "but having this fancy calibrated stick looks more professional. Too bad it's in metric, though," he added. "I can't read metric, so I have to convert everything to imperial. I bought this stick in metres because when my son takes over the business, he won't know how to read inches and feet."

When Joe Zeffer, the backhoe operator, had dug a hole near the stake that looked about deep enough, I went over to it and held the stick in the hole while Werner squinted for the surveyor's tape through the level. "Nope, another yard," he called. Joe went back to digging. Once, Joe got out of his machine to check the depth of the hole himself. "I'm only at 3.46 metres," he said. "I've got to go down another 6 or 8 inches," and he climbed back into the cab. His mixture of metric and imperial measurements reminded me of Brian Van Flught's earlier description of the standard iron bar: "25 centimetres square and 4 feet long."

It was a smallish backhoe, a John Deere 490D, on loan from Sittler Excavation Ltd., a company based in Elmira, Ontario, a village a few miles from Waterloo. The machine had a ditching bucket that Joe said held "about a yard and a quarter," meaning 1¼ cubic yards of dirt. A ditching bucket is like a standard bucket, except that it doesn't have teeth; it scrapes along making a smooth bottom, which is what you want in a ditch, and works excellently in sand or soft soil but not very effectively in the hard, compacted clay we were in here. The biblical injunction to build your house on a rock was not written by a backhoe operator. A backhoe operator would have gone for sand every time. Still, the work moved along. Joe rimmed the excavation with piles of dirt, manipulating the bucket with the ease of a child playing with a spoon. Gradually, the hole began to assume the shape of a foundation.

When one corner of the hole was exactly 7 feet 8 inches below the top of the stake, Werner took a length of 2-by-2 out of his truck and fashioned it into a vertical free-standing pole, which he propped up at the corner like a kind of branchless Christmas tree. Then he brought the level down into the hole, set it up in a different corner, and we sighted a line on the 2-by-2. We now had a line on the 2-by-2 that was exactly the proper depth; all we had to do was keep moving the 2-by-2 around in the hole and squinting at it through the level until no matter where we put it, the cross hairs in the level hit the line on the 2-by-2. If the cross hairs hit the 2-by-2 above the line, the hole was too deep. If they hit it below the line, then Joe had to scrape deeper with his toothless bucket.

Sometimes he scraped so hard that the near tread under the backhoe's cab lifted entirely off the ground, and the machine, with Joe in it looking unconcerned, was balanced precariously on the edge of the far tread and the tip of the bucket. Sometimes he had to do this to scrape off less than an inch of clay. It seemed simple. It took seven hours. When he finally shut off his machine at 5 o'clock, Lot 92 was a rim of piled clay at the centre of which was a square hole, 7 feet 8 inches deep at the Dream Home side, tapering off to about 4 feet at the other side because of the slope of the lot. The walls of the hole were 5 feet beyond where the walls of the house would be, and the bottom of the hole was a smooth, flat surface of light brown compacted clay, so compacted that Werner could barely drive a stake into it with a sledgehammer. From a house builder's point of view, it was a good compromise between the sand preferred by the digger operator and the rock recommended by the Bible.

□ □ □ □

Tim announced the Advanced Houses program in a brochure sent out to members of the Canadian Home Builders' Association in April 1991. "An Advanced House," he wrote, "is an exploration beyond conventional design and construction practices." Six houses, he said, would be built by the winners of a design competition; winning designs would be of "the most energy-efficient and environmentally responsible homes in Canada." The program, he hoped, would "lead the way toward a new generation of homes built on a solid foundation of energy-efficient and environmentally appropriate construction principles." Bringing together all the different components of the building industry – designers, architects, builders, renovators, manufacturers, researchers and utilities – would "help to determine which innovative systems, approaches and products are most suitable for adoption by the building industry. Suitable new concepts," he continued, "may also be considered for integration into an updated R-2000 technical standard."

Like Brampton House, the new Advanced Houses would be twice as energy-efficient as R-2000 houses; but again, the way each house achieved that goal would be up to individual designers. This time, however, Tim specified areas in which the energy reductions were to be concentrated: "50 percent less for space heating, 50 percent less for lighting and 50 percent less for outdoor electrical usage." The new houses also had to use a lot less water than a conventional house, use recycled or recyclable building materials (locally manufactured, where possible, to cut down on the energy involved in transportation) and incorporate on-site waste-management schemes to reduce the amount of construction waste needing to be hauled to a landfill site.

The brochure began with a brief statement of the program's objectives: "1. To conduct regional field trials of new concepts, technologies and products to assess their overall performance and their suitability for adoption by mainstream builders and for their inclusion in the R-2000 technical criteria." The operative word here is "regional." No single energy-conscious idea or gizmo will work equally well in every region of the country: factors such as climate and material availability will change a product's usefulness from place to place. At present, however, very similar standards are applied in large-scale housing developments across the country. And, although R-2000 building practices made sense to the small custom builder who had time to double-caulk every seam in a vapour barrier, Tim hoped that the Advanced Houses program would devise ways to show large developers that environmentally appropriate houses don't have to cost more or look worse than conventional houses.

"2. To encourage the adoption of successful, new, energy-efficient and environmentally appropriate technologies by the residential construction industry." Low-energy technologies designed and developed in Canada—like the heat-recovery-and-ventilation (HRV) system built for Saskatchewan House in 1977—have become major industries: Canadian companies now manufacture and export 50,000 HRV units a year. Tim wanted to see similar prototypes in other low-energy areas field-tested in this new series of Advanced Houses.

"3. To provide a technical basis for the future development of a new, updated R-2000 standard in reaction to advances in housing technology and emerging environmental concerns." The National Building Code (NBC) is due for revision in 1995, and EMR wants to include most of the R-2000 energy targets in it: the Advanced Houses program would be a good opportunity to test and even upgrade those specifications.

The Advanced Houses program would judge entries from designers and builders according to how well their submissions met a series of stringent technical requirements in nine categories: airtightness, ventilation systems, combustion equipment, energy performance, water usage, recycling, ecomanagement, indoor air quality and monitoring.

Airtightness. The requirements specified a maximum level of air leakage through the building envelope of 1.5 air changes per hour (AC/hr). The proposed air-change level would be analyzed by EMR's HOT-2000 computer program, which predicts how much energy will be consumed in heating the house, a direct offshoot of the house's airtightness. The lower the space-heating requirements, the better chance the design would have of being accepted. Thus, although 1.5 AC/hr represented no advance on R-2000 requirements, competitors were encouraged to beat that level because the HOT-2000 run would award points on the designers' *proposed* air-change level, not on a computer prediction.

Ventilation Systems. Each house had to have a mechanical ventilation system to provide continuous exhaust of used air and intake of fresh air at rates that met or exceeded levels recommended by the Canadian Standards Association (CSA). HRVs had to be situated no more than 2 metres from an outside wall, and the duct that carried the cold air into

GREENHOME DECISION MATRIX

During the design process, the Greenhome team assessed each building material according to criteria outlined in the "decision matrix." Only those materials that met these rigid environmental standards were used.

MINIMUM ENERGY REQUIRED TO BUILD AND OPERATE THE HOUSE
For each material or system in the house, how much energy is required to:
produce the product (embodied energy)?
transport the product to the job site (embodied energy)?
install the product (embodied energy)?
operate an energy-consuming device (energy efficiency)?
replace the product at the end of its useful life (durability)?
dispose of the product at the end of its useful life (reusability, recyclability)?

MINIMUM NEGATIVE IMPACT ON THE GLOBAL ENVIRONMENT
Are toxic chemicals released during manufacturing?
Are acid gases, greenhouse gases or CFCs released during manufacturing?
Are other air or water pollutants released during manufacturing?
Are long distances required for transportation?
Is the product produced from renewable resources?
Does the product contain recycled materials?
Is the product recyclable at the end of its useful life?

MINIMUM NEGATIVE IMPACT ON THE LOCAL ENVIRONMENT
Minimum water use for water-consuming devices?
Are toxic or other chemicals given off during the life of the product?
Can scrap products be returned for reuse or recycling?
Will any component associated with the product be land-filled?
Are chemicals required to be applied to the product?

MINIMUM NEGATIVE IMPACT ON INDOOR AIR QUALITY
Are chemicals given off during the operation/life of the product?
Will the growth of microorganisms be supported during the life of the product?

OTHER CONSIDERATIONS
Is the product locally produced?
Does the product have Canadian content?
Does the product have the required code approvals?
Can the product be easily adapted by the building industry?

ENVIRONMENTAL COMPARISONS

Traditionally, building components are selected on the basis of price and performance. But the environmental issues facing the world today suggest that a new approach is required. At the Waterloo Greenhome, the environmental consequences associated with the production, installation, use and disposal of building components were identified in the selection process. Here are two examples of the team's assessments.

DOORS

ENVIRONMENTAL IMPACT OF:	WOOD	STEEL	FIBREGLASS
Resource Depletion	Medium	Low	Low
Manufacturing Emissions	Low	Medium	Low
Embodied Energy	Low	Medium	Medium
Energy During Use	High	Low	Medium
Ozone Depletion	Low	High	Low
Emissions During Use	Low	Low	Low
Disposal	Low	Medium	Medium

INSULATIONS

ENVIRONMENTAL IMPACT OF:	FIBREGLASS	PHENOLIC FOAM	CELLULOSE
Resource Depletion	Medium	Medium	Low
Manufacturing Emissions	Low	Low	Low
Embodied Energy	Medium	Medium	Low
Energy During Use	Medium	Low	Medium
Ozone Depletion	Low	Medium	Low
Emissions During Use	Low	Low	Low
Disposal	Low	Low	Low

the house and the warmed air out of the house had to be insulated to a minimum value of R-3, an improvement on the existing requirements for R-2000 houses.

Combustion Equipment. All furnaces and water heaters had to be direct-vented to the outside, which meant they had to draw their combustion air from outside the house rather than from within it, and they had to exhaust their gases directly outside rather than into the house. Wood-stoves and fireplaces also had to be direct-vented, and ducts supplying these devices had to be insulated to R-20.

Energy Performance. This was a big one, since it included specifications for the houses' total energy-consumption levels, which is the sum of space-heating, space-cooling, water-heating, light, appliance and outdoor electricity requirements. The total performance target was set at half the R-2000 levels. The annual space-heating consumption would be determined by the HOT-2000 computer program, using an equation that took into consideration such factors as the houses' volumes, their surface-to-volume ratios, number of occupants, kind of heating fuel and the annual number of heating degree-days in their localities.

"The houses should incorporate passive design features," the list of technical requirements suggested, "such as window shading, window overhangs, thermal mass, et cetera, to limit or remove the need for mechanical cooling." But each designer was free to achieve a low-performance target as best he or she could. Another suggestion was aimed at eliminating conventional air conditioners: "If vapour compression refrigeration is used, only those cooling devices using refrigerants with an ozone-depletion factor less than or equal to 0.05 are permissible." This meant, basically, no Freon or CFC-blown insulation in the air conditioners, refrigerators or freezers.

Water heaters had to provide hot water at a minimum temperature of 50 degrees C and be more than 75 percent efficient. Preheat tanks, solar water heaters, heat pumps and heat exchangers were encouraged.

To the call for proposals, Tim appended a few charts and graphs showing the difference between conventional house performance and the targets he expected the Advanced Houses to meet. The total annual electrical consumption of appliances in a typical home in which all appliances conform to EnerGuide ratings is about 7,100 kilowatt hours per year (kWh/yr). For consumption figures, see chart on opposite page.

The electrical-energy target for an Advanced House was set at 3,838 kWh/yr, so some dramatic cuts had to be made. Builders were to supply major appliances, including refrigerators, stoves, clothes washers and dryers and dishwashers, each with a specified minimum capacity (that is, they couldn't cheat by installing a bar fridge or an apartment-sized washer/spin-dryer). The total consumption for these appliances, according to the above chart, would be only 2,556 kWh/yr, leaving about 1,200 kWh for extras. Some choices were obvious—if you put in a water-bed heater, you couldn't have a freezer—others were not. As we shall see.

Water Usage. The use of low-flow toilets was encouraged. An average toilet uses up to 20 litres of water per flush; Tim decreed that in an Advanced House, "the amount of water per flush shall not exceed 7 litres." Shower heads were restricted to 10 litres per minute; all sink faucets had to have aerators; landscaping had to use "native vegetation

ENERGY CONSUMPTION

APPLIANCE	KWH/YR
Refrigerator	240
Stove	780
Clothes washer	144
Dryer	1,140
Dishwasher	252
TV	312
Furnace-fan motor (continuous)	1,512
HRV	631
Central vacuum	96
Electric clocks (2)	96
Hair dryer	90
Iron	90
VCR	54
Water-bed heater	1,152
Freezer	504
TOTAL	7,093

with low water requirements" and have minimal lawn areas. "The goal is a 50 percent reduction in outdoor water use relative to common practice in the area in which the house is to be built."

Recycling. Houses had to provide areas for the storage of recycling bins for paper, metal and glass, as well as an indoor composter with at least a 10-litre capacity and an outdoor composter with a 150-litre capacity.

Ecomanagement. This category dealt with the houses' impact on the local environment. Recycled materials were to be used for construction whenever possible, and the amount of construction waste trucked to a landfill site had to be kept to an absolute minimum. "For instance, waste gypsum-board products shall either be sent to a recycling plant or be used as thermal mass within the building." Scrap drywall can be placed inside the wall and floor cavities, for example, which not only saves landfill sites but also contributes to soundproofing between rooms.

Materials were given an environmental rating called an ozone-depletion factor, based on the use of chlorofluorocarbons (CFCs) in their manufacture. No material could be used that had an ozone-depletion factor greater than 0.05. This meant that any material used in the house which had been made with CFCs had to save more energy than was used in its manufacture. For instance, 2 inches of foam insulation (R-12) has an ozone-depletion factor of 1 and therefore was unacceptable; 4 inches of foam insulation (R-22) has a factor of 0.05, because the amount of energy needed to make the foam is more than offset by the amount of energy it saves, so it could be used.

One of the major thrusts of the Advanced Houses program was its attack on space-heating bills. In a cold climate, space heating uses up as much as 30 percent of the energy available to us. The R-2000 program tried to reduce that percentage. Houses built in 1975 to NBC specifications used an average of 217 kWh of electricity per square metre of floor space. By 1983, with advances in such things as insulation, that figure had been reduced to 167 kWh/m² in houses conforming to standards in *Measures for Energy Conservation in New Buildings* (simply called *Measures*), a nonmandatory companion to the NBC. R-2000 houses built before 1985 cut the 1975 figure in half: 104 kWh/m² per year. Brampton House, completed in 1989, tests in at 123 kWh/m². The target for the Advanced Houses was 52 kWh/m².

How does that compare with an average house? I can use our own as an example. Built in 1917, it's a 2½-storey, six-bedroom house with a heated basement and a 12-by-25-foot addition on the back, giving a total floor area of about 70 m². We have a low-efficiency oil furnace, but high-efficiency refrigerator and freezer, and since moving in three years ago, we have added such energy-saving features as insulated metal doors front and back and double-glazed, gas-filled, low-E-coated windows downstairs; we also superinsulated the addition, sealed all the windows in the fall, stuffed insulation around the floor joists at grade level in the basement and generally tightened up the house to make it relatively comfortable in winter. In 1991, we used a total of 8,847 kWh of electricity and burned 3,295 litres of furnace oil. Converting the oil units to electricity units (1 litre of oil is the equivalent of 10.75 kWh of electricity), we used a total of 35,421 kWh, or 500 kWh/m², roughly five times more than an R-2000 house and 10 times the target set by Tim for the Advanced Houses program. Crudely put, the energy our house consumed in 1991 cost us $2,061; the energy for a house the same size, built to Tim's technical specifications, would cost about $220.

□ □ □ □ □

Tim decided that the best way to reach builders was through the national network of homebuilders' associations, so shortly before the official announcement of the Advanced Houses program, he organized a "warning meeting" of representatives from each of the provincial homebuilders' associations. Since Enermodal had been monitoring the energy performance of Brampton House, Steve was invited to the meeting to give some technical input into what an Advanced House was.

At the time, Steve assumed that because Ontario already had Brampton House, preference for the six new houses would be given to proposals from outside the province. When he asked Tim directly whether another Ontario house would be considered, he says, "Tim gave the usual response, saying, 'Well, if it's good, of course it will be considered.' But I didn't think there was much of a chance."

Still, Steve was interested. One of the other invitees was Nancy Cowan, who worked for Union Gas, the utility that serves most of southwestern Ontario, including Kitchener-Waterloo. Steve knew that one of the most common criticisms of Brampton House was its reliance on electricity. There was a big push in the low-energy industry to switch to natural gas, so he asked Nancy whether Union Gas was interested in participating in an Advanced House.

"Very much so," she replied. "That's why I'm here." Back in Waterloo, Steve called a meeting at the Enermodal offices and invited Nancy from Union Gas, Mike Jacobs and Carolyn Kinsman from Ontario Hydro, Allen Jenkins from the Ontario Ministry of Energy and Tony Krimmer, head of Waterloo's building inspection department. Also at the meeting was Steve's colleague at Enermodal, John Kokko. And Elizabeth White. "She started off with a pep talk on the Brampton House," says Steve, "and then I said there was a reasonable chance that Tim would fund another Ontario Advanced House, but to get to build it would be a tough fight, because another Ontario house would have to be much, much better than Brampton House. Then I went around the table and asked, 'Are we interested?' And everyone said, 'Yeah.' "

Then Steve went on to interpret Tim's ground rules. "They were very simple: he wanted a house that would use half the energy of an R-2000 house, and he didn't care how we got there. We could put our own regional context to it. The only thing Tim added from the Brampton House was the environmental side—don't use any CFCs, use eco-local products, have inside storage for a Blue Box." Those at the meeting thought a Waterloo house could go well beyond those guidelines, provided they could raise the money.

"At that time, Tim said he had a budget of $1.5 million and they wanted to build six houses, so those of us who were very clever with math figured they were prepared to spend $250,000 a house. This was for the incremental construction costs. You know, you want to put in a low-water-use toilet, for example, and they cost $1,000 while a standard toilet costs $200; they would cover the $800 difference. They did not want to pay for the basic construction cost and plans. They were working on a leverage of 3 to 1: for every $1 they put in, they wanted to see $3 of other people's money. We had to come up with that $3 ourselves, in the form of either donated services or products or time or cash or however we wanted to count it."

From their work on Brampton House, Steve and Elizabeth knew they could meet the technical requirements and even go beyond them. "We saw instantly that the financial challenge was going to be harder than the technical challenge."

Tony Krimmer was a big help. As a longtime building inspector, he knew every builder in the Kitchener-Waterloo area and could easily separate the good from the gimcrack. He was also involved in the educational arm of the Ontario Building Officials Association, teaching courses in Ontario Building Code interpretation—engineerese as a second language. Tony was the first person in Canada to open his course to builders and architects as well as to their future nemeses, building inspectors. One of his graduates was Werner Reiter, a small homebuilder who had also been trained as an R-2000 builder. Richard Reichard, a successful architect whose offices were just across Columbia Street from Enermodal, had also taken Tony's course. Tony asked Werner and Richard if they were interested in climbing aboard, and they both said yes.

Steve turned his attention to raising the capital. "We just basically got on the phone and started talking to people. In the end," he says, "it wasn't that difficult. I knew that it was a bit like getting investors interested in a new invention—once you get the first guy to wade in, the others will leap. In our case, Tim was that first guy. If our design won the competition, then EMR was in for a quarter of the construction cost."

Following Tim's strategy—or rather, anticipating it, since the official EMR call for proposals hadn't gone out yet—the first group they approached was the Kitchener-Waterloo Home Builders Association (KWHBA), whose president was Ian Cook. "Our timing was lucky," says Steve. "Cambridge homebuilders had just done a sort of charity project—they'd got all their members together to come and work on weekends to build an 'affordable house' in Cambridge, and then they donated it to charity. They knew that the stereotype of a builder is of a money-grubbing guy who puts up junk and sells it for as much as he can so that he can run off to Florida. These homebuilders were saying, 'We're not really like that, we want to put something back into the community.' They got a lot of press out of it, and the Kitchener-Waterloo homebuilders were looking for something to do too. So when we went to see them, they were ripe for a proposal."

The homebuilders and Enermodal agreed to split the project in two. "We told them that there would be the basic construction costs, and then there would be what we called the soft costs. Some of them aren't too soft, mind you, but they are things that are extraneous to the basic building – design fees, for example: builders are used to just pulling a stock plan off a shelf and building 50 of them. Well obviously, if you're going to build the most innovative house in the country, you've got to spend a little time thinking about it first. And the monitoring: there's no point building the most advanced house in the country if you can't quantify that it works."

After a long series of talks, Steve and the KWHBA, via Ian, came to an agreement. Steve estimated that the total cost of the project would be "basically $760,000," including the land ($70,000), the construction costs and the incremental costs that would be covered by EMR. Of that, $200,000 would be borne by the KWHBA, which would own the house when it was finished and sell it for whatever the market would bear. "They went to their members and said, 'Build this house, supply the materials, and you get paid when we get paid.' So the plumber goes in and puts the piping in, and it may be a year or two before they sell the house and he sees any money. And he'll get paid on a prorated basis. If everything works out fine and they sell it for $200,000, the plumber bills them for $1,000 and he gets paid $1,000. If the house sells for only $150,000, then the plumber gets paid only $750. On the other hand, if the house sells for more than $200,000, the plumber gets his $1,000 and the rest will go to charity. No one is looking to make a profit out of this, which I felt was very nice; again, it's the whole idea of putting something back into the community."

Armed with this agreement and similar indications of support from City Hall, Waterloo Hydro and Union Gas, Steve was able to sit down with his colleagues at Enermodal and begin to work out their submission to EMR. He wasn't naïve about the homebuilders. He knew that there was a lot of interest in the project from them, but he also knew that individual members would be donating their time and materials in the expectation of getting paid in full when the house was sold. This meant pressure on him to design a house that would be sellable. He would have to be extremely diplomatic in order to get his design principles across without compromising the environmental integrity of the project.

In the end, he decided to design the house that he had to design and thrash out the details later. He was tremendously excited. He was embarking on the project of a lifetime. Not only had he been given the green light to dream up the most energy-efficient house in the world, but he was also setting the standards for low-energy houses for the future. To do it, he knew he would have to change the way everyone, not just builders, thought about houses.

THE DIFFICULTY WITH MECHANICALLY EXTRACTABLE
ENERGY IS THAT SO FAR WE HAVE BEEN UNABLE TO MAKE IT
AVAILABLE WITHOUT SERIOUS GEOLOGICAL AND
ECOLOGICAL DAMAGE OR TO EFFECTIVELY RESTRAIN ITS USE
OR TO USE OR EVEN NEUTRALIZE ITS WASTE.
— WENDELL BERRY, "THE USE OF ENERGY"

Over a beer at the local pub in Carp, Ontario, where he lives, Oliver Drerup is telling me a story about changing the way we think. "It's a staggering story, if I do say so myself," he says. "It's about a man named Bob Berkebile. Are you familiar with the Kansas City Hyatt Regency?"

I wasn't, but Oliver quickly filled me in. On July 17, 1981, four floors of elevated concrete walkways, or skybridges, that ringed the atrium lobby of the Hyatt Regency Hotel in Kansas City collapsed during a Friday night dance. There were 1,500 people in the lobby; 114 of them were crushed to death, and hundreds of others were injured. Investigators for the U.S. National Bureau of Standards called it "the most devastating structural collapse ever to take place in the United States." At the hearing, responsibility for the disaster and $3 billion in lawsuits were levelled at the hotel's steel fabricator, the structural engineer who okayed the plans, the general contractor who carried them out and the architect who designed the building in the first place.

"Berkebile was the architect," says Oliver. "When the thing crashed, the rescue workers called him in to assist them in the rescue operations. They were lifting huge pieces of concrete off dead people. Well, you can imagine how he felt, seeing bodies coming out of a building he designed. As it turned out, the crash was caused by a plan change after the original design work had been done, not by a flaw in the architectural plan, so it wasn't Berkebile's fault, but he didn't find that out until six years later. When I met him last year, he told me that his life had been turned upside down. He had gained an insight into the consequences of architecture.

56 "After he was cleared, Berkebile designed another building, a high-

rise office tower, and he specifically stated that no wood veneer that came from rainforest lumber was to be used in the boardroom. No teak, no mahogany, nothing. He was a changed man. He had become concerned about human life, about the way we are treating the environment. Anyway, when the building was finished, he was invited to the opening ceremonies. They held a cocktail party in the boardroom, and guess what: the whole room was done in mahogany veneer. Berkebile was furious; he could hardly keep himself from shaking. As part of the ceremony, the company that had made the panelling presented the president of the company with a photograph of the tree from which the veneer had come. By now, Berkebile was beside himself; he didn't know what to do, so he turned to a guy standing next to him, and he said, 'I don't believe it; I told these guys not to cut down a rainforest tree to make this veneer, and look what they did.'

"Well, the guy standing next to him says, 'You know, it's funny, because my company supplied the aluminum for the exterior of this building, all 37 storeys, and to get at the bauxite we use for our smelting process, we cut down acres and acres of rainforest in the Amazon valley every year. And here you are getting upset about one tree.' "

Reconsidering traditional building practices is not entirely new: consider a book published in 1932 by American writer Stuart Chase entitled *The Tragedy of Waste*, the earliest study I know of that traces the enormous impact of what has come to be called "planned obsolescence." Chase considered various sectors of the North American economy, including housing, and concluded that the most formidable enemy of economic progress was industrial waste. In his book, he charted hundreds of examples of waste in American industry that contributed to the nation's "illth," which is the opposite of wealth.

One of his examples was waste in architecture. He began by citing F.L. Ackerman, "an architect who has given wide study to housing wastes in New York City." Ackerman was shocked to discover that "city apartments are now being erected as short-term investments, to be torn down and rebuilt as land values shift under them," and went on to disclose that "all the plumbing in New York City has to be replaced every eight years." Since it took twice as many plumbers to replace pipes as it did to install them, "we now require three times as many plumbers to install and reinstall as would be the case if buildings were made durable in the first place."

Chase also quoted another New York architect, Ernest Flagg, who railed against shoddy construction practices: "The ordinary small house," said Flagg, "is almost as bad a liar as the tall building. . . . Inside the house, expensive shams and concealments abound, and the result is thought to be 'artistic.' Door frames are not real but shams, the real frame being covered with casing or trim which requires six times as much inflammable material and six times as much time and labour to make and set as would be necessary if the true frame were made presentable and shown." Houses, in other words, were becoming replaceable. In 1925, the American Construction Council reported that building methods in most American cities were "deplorable" and that many new small houses "will be worthless within 10 years."

Chase was concerned about the economic impact of illth, but he was also doing the groundwork for an appreciation of the amount of energy required to produce a given material and factoring that into the total energy budget of the product the material is used for. For Chase, the impact of wasted energy was economic: paying three plumbers to do the work of one was simply needless expense. For more recent observers, the effect of wasted energy is environmental: those three plumbers are laying twice the amount of pipe they should be, which requires twice the amount of energy to produce, half of which is entirely wasted.

In his 1977 book *Architecture and Energy*, Richard G. Stein estimated that "the manufacture and installation of building components is responsible for about 8 percent of all energy for all purposes" in the United States. In 1970, that represented 5.75 million billion BTUs of energy, the equivalent of 38 billion gallons of oil or an entire year's output of 385 thousand-megawatt electrical generating plants (and there are only 750 of them in the United States). Today, when our total energy consumption has leaped to more than twice what it was before OAPEC (Canada produced 205 million MWh of electricity in 1970 and 465 million in 1990), the amount of energy devoured by the building industry, 40 percent of which is in the building of houses, is even greater: Philip Steadman, author of a recent report to the Academy of Natural Sciences entitled *Energy, Environment, and Building*, places "the amount of energy which is consumed in residential and commercial buildings for all purposes at around 33 percent of total U.S. energy consumption."

Just how much of that energy is wasted has been shown by the success of the conservation movement: if we conserve electricity, for example, and the electrical companies record a net decrease in the amount we use without a corresponding decrease in our economic well-being, then the amount of electricity not used must have been wasted before we stopped using it. Here are some case histories taken from Lawrence Solomon's book *The Conserver Solution*, published in 1978: in 1974, Carleton University, in Ottawa, spent $20,000 on double-glazed windows and

CONSTRUCTION WASTE

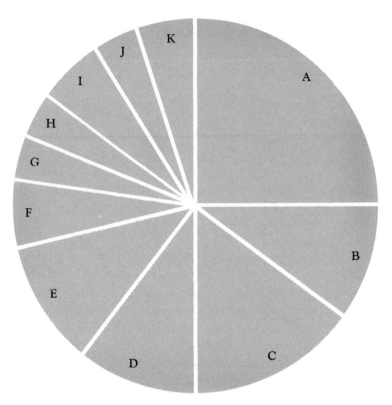

	WASTE PRODUCT	PERCENTAGE VOLUME	AVERAGE WEIGHT/HOUSE
A	Dimensional Lumber	25.0%	0.845 tonne
B	Manufactured Wood	10.0%	0.424 tonne
C	Drywall	15.0%	n/a
D	Masonry and Tile	12.0%	1.00 tonne
E	Corrugated Cardboard	10.0%	0.066 tonne
F	Asphalt	6.0%	
G	Metal Wastes	4.0%	
H	Plastic and Foam	4.0%	
I	Fibreglass	5.0%	
J	Other Packaging	4.0%	
K	Other Wastes	5.0%	

instituted such conservation measures as lowering thermostats in class-rooms and offices at night and saved $600,000 a year in electricity bills. In 1977, the federal government imposed similar measures on its buildings and saved $30 million. Dominion Foundries and Steel, of Hamilton, Ontario, simply began turning off the lights in its headquarters building on weekends and saved $75,000 a year. Clearly all that *money* had been wasted before: but more important to the environment, all that *energy* had been wasted before.

Creating material requires energy, so wasting material wastes the energy embodied in it. Throwing away aluminum wastes not only rainforest but also the fuel burned in the ship that brought the bauxite to the smelter and the fuel in the trucks that hauled the aluminum from the smelter to the supplier, from there to the job site and from the job site to the landfill. In 1989, *Time* magazine reported that "Americans throw away enough aluminum to rebuild the entire U.S. commercial airline fleet every three months." Construction waste now accounts for 15 percent of all material tipped into Canadian landfill sites; the average new house in Canada generates 2½ tons of construction waste.

We must begin to make intelligent choices, not only about the fuel we burn to heat our houses but also about the materials we use to build them, and most of those choices involve waste. Look at electricity, for example. During the 1970s, the big push was from oil heating to electricity. In 1978, 44 percent of all Canadian houses were heated with oil and 17 percent with electricity; today, the figures are almost reversed.

But heating houses with electricity hasn't been the miracle cure we thought it would be: about half of all the electricity produced in Ontario comes from nuclear generating plants, and half of the rest comes from the combustion of fossil fuels. A recent report submitted to Ontario Hydro estimated that the average electrical generating plant, whether nuclear, hydro or fossil-fuel, operates at less than 35 percent efficiency and that a further 10 percent of the electricity generated is lost to resistance in transmission lines. This means that every time enough fuel is consumed to produce 100 kWh of electricity, only about 20 kWh are actually available to the consumer. The rest is waste. Our use of electricity makes New York City's plumbing look positively anal retentive.

This was bad enough, from an environmental point of view, at a time when most electricity was produced at Ontario Hydro's coal-burning facilities: coal requires a huge and ever-increasing amount of primary energy just to wrest it out of the ground; changing it into coke, the form in which it is burned in the generating plants, spews tons of smoke and gases into the atmosphere, contributing to air pollution and to global warming; and the burning of coke at the generating station sends up more tons of sulphur dioxide, nitrogen oxides, methane and other prime ingredients of acid rain (Ontario Hydro is the second-largest source of acid rain in the province, after INCO). But with half of Ontario's electricity being produced in nuclear-powered generators, the environmental impact of electricity usage—and wastage—is even greater.

The second consideration is economic: it costs more to produce heat from electricity than it does to produce it directly from a fossil fuel—14 times more than oil and gas when the electricity comes from a coal-burning generator and 23 times more when it comes from a CANDU reactor. As John Kokko puts it, "Electricity is too precious to waste on something as crude as space heating. There are better uses for it—running computers, for instance—and more efficient ways to heat a house."

Steve likes playing with numbers. When he devised a way to translate energy statistics into more tangible analogues, he came up with the "environmental costs" of fuel burning: at the 1990 annual meeting of the Solar Energy Society of Canada Inc. (SESCI), he presented his findings in a paper in which he assigned a dollar value to each form of energy source in terms of its environmental cost. "All energy sources are not equal," he said. "The use of some energy sources causes severe environmental damage, whereas others are relatively benign." The burning of coal, for example, contributes to the greenhouse effect and global warming, acid rain and ozone depletion—a heavy environmental cost—and although the replacement of coal by renewable energy sources does not eliminate environmental costs, it certainly reduces them.

John mentioned one day that even hydroelectric plants contribute to environmental degradation, because the huge artificial reservoirs of water required to run the turbines destroy wildlife habitat and kill trees that would otherwise turn carbon dioxide into oxygen. But compared with coal or uranium, the environmental cost of water power is small. At the SESCI conference, Steve cited a 1988 study which found that even the manufacture of a high-tech item like a photovoltaic cell has 200 times less environmental impact than the construction and operation of a fossil-fuel-burning electrical generating plant.

Using those and other available figures, such as the cost of acid-rain cleanups and the loss of farmland (Steve is not the only person looking at these things), he came up with a chart which showed that the environmental cost of producing electricity from coal is twice that of producing it in a nuclear generator and nearly three times that of producing it from natural gas. The environmental cost of materials chosen for the construction of the Greenhome would be given serious consideration.

The Greenhome incorporates all the features of earlier low-energy houses: R-34 wall insulation, south-facing triple-glazed superwindows and a supertight air/vapour barrier. But it also uses recycled and environment-friendly materials.

With a clear sense of what they wanted to do with the Greenhome, Steve and his colleagues drew up their submission to the Advanced Houses program. It was submitted in January 1992. In a statement of the Greenhome's aims, Steve pushed all the "hot buttons" on Tim's switchboard: "The primary goal of the project is to demonstrate to the building community and the public that new homes can be built and operated in a manner that has significantly less impact on the environment and is acceptable to the home-buying public." However, the submission emphasized that using energy-efficiency systems that have already been accepted in the marketplace or, conversely, employing highly experimental or radically innovative features that will never be fully adopted by the industry would not get them anywhere. "The design of the Greenhome must be made up of features that may be considered nonconventional but are commercial or near-commercial and could easily be adopted by the building community."

Without going into too many specifics (materials and systems still had to be selected), the proposal gave the general outline of their plan. The Greenhome would be a single-family unit in the raised-bungalow style, since that's what everyone is building these days, with bedrooms and a bathroom in the basement "so that it could be used as a separate apartment for grandparents, guests or extra space for a growing family." This would be received favourably by the city of Waterloo, which was encouraging suburban planners to make houses "duplexable." The shell would be superinsulated and airtight; many of the materials used would be recycled—including remanufactured wood siding, drywall with recycled gypsum, recycled vinyl flooring, laminated lumber made of recycled wood waste, and so on. No construction waste would go to a landfill site; all of it would be sold or given to a recycling station. And they would come in under EMR's energy-consumption target of one-half R-2000. Total estimated cost of the project, including donations from builders and suppliers and the incremental costs covered by EMR: $757,250.

Steve put the 68-page proposal in the mail on January 3. He still didn't know whether Tim would go for another house in Ontario, but he had fulfilled the criteria. All he could do now was wait.

□ □ □ □ □ □ □

In the proposal, Steve named the project The Waterloo Green Home. In the October-November 1990 issue of *Solplan Review*, which had arrived at the Enermodal office while Steve and the others were drawing up their proposal, was an article by Vancouver architect Richard Kadulski in which he called for a new way of thinking about house building. "Reducing the amount of waste generated during construction is only the start," he said. "We must look at construction with a new viewpoint. We must look at the kind of buildings we build, the materials, the operating costs. In the absence of another term," he added, "I'll call it 'The Green Home.' "

Kadulski, who is now the project manager of the British Columbia Advanced House, outlined a series of principles to define his concept. A green home, he said, should be energy-independent. "It is possible to build homes in most parts of Canada that require no special heating systems." A house, said Kadulski, should not pose a health hazard to its occupants: materials should not gas off or otherwise contribute to indoor air pollution; building and décor materials should include more natural products, like wood and cloth, and fewer synthetics, like nylon and plastic. A house should not be larger than the space requirements of the people who live in it. "The most energy-efficient mansion sprawling over thousands of square feet to house a couple with one child is not efficient, nor is it responsible."

In Kadulski's view, the word "responsible" is the key, and it carries a double indemnity. We are, in both senses of the word, responsible for the environment: we got it this way, and it's our responsibility to clean it up. Making our houses environmentally responsible requires nothing less than a complete shift in the way we think about our surroundings. Turning from the kind of houses we've been building in the past to the kind of houses that Kadulski envisions will be as big a revolution as the one that turned many of us from building the uninsulated, synthetics-filled gas guzzlers of pre-OAPEC days to building the low-energy, superinsulated, integrated systems of the 1970s and 1980s.

Houses, says Kadulski, like industry, like governments, "must have a low environmental impact." This goes well beyond installing low-flush toilets and high-efficiency gas furnaces after the house is built. It means looking at the total environmental impact of a nail before you drive it into a piece of wood. It means rethinking everything. "Materials used in construction should have little embodied energy or not be the end result of toxic manufacturing processes."

Kadulski was heralding a new age in house construction. "We must simply consider how all of these go together," he concluded. "Builders that get involved now will be on the leading edge of environmentally friendly, healthy homes."

And that's exactly where Tim Mayo wanted the Canadian construction industry to be.

THE FOUNDATION

□

THE BEST FRIEND OF THE ARCHITECT IS THE PENCIL IN THE
DRAWING ROOM AND THE SLEDGEHAMMER ON THE JOB.
—FRANK LLOYD WRIGHT

In house building, things move either very quickly or not at all. During the last week in September, they weren't moving at all. Perhaps it was the fall weather, but the air seemed charged with readiness: clear and crisp, the light slanting across the lot and filling the excavation with warmth. The lot had been surveyed and staked, the hole had been dug, and the footings for the foundation were in place. As yet, however, there was no foundation for them to support.

Delays are usually built into the construction cycle, but that doesn't stop project managers from kicking against them anyway. Werner brooded around the lot on Tuesday, muttering about frost and rain and snow and other factors that wouldn't obtain on Wednesday but might on Monday. The fence between the Dream Home and our lot looked poised to throw itself into the hole at any moment, and Werner convinced himself that one good rain would bring the whole thing down in some Vesuvian mudslide, along with the garage and the brand-new Ford Mustang that was in it (second prize in the Rotary draw). Rain would also turn the job site into a quagmire. There was some consolation to be taken from the fact that the foundation system didn't depend on weather conditions at all, but Werner wasn't taking it. He's a builder; any time not spent actually building is time wasted, and in this project at least, waste was supposed to be eliminated. Besides, he didn't like the foundation system chosen by the designers very much, even though he'd never seen it used before. Scratch that: he didn't like it *because* he'd never seen it used before. "Show me some engineering data," he had said at a design meeting in February. "Unless it's been used in the field for a year, I won't approve it."

A number of innovative foundation systems were discussed at the design meetings held in the Waterloo boardroom of the Greenhome's architects, Snider Reichard & March. Richard Reichard, who was chiefly responsible for the architectural aspects of the house, usually sat at the end of the huge board table beside Ed Bordeaux, the architect who would actually draw up the plans. John Kokko unofficially chaired the meetings; Elizabeth White was back as a consultant; Ian Cook represented the Kitchener-Waterloo Home Builders Association; and Tony Krimmer represented the City of Waterloo. Tony had a big green-and-white binder on the table in front of him, the 1990 Ontario Building Code, and he looked as though he was prepared to use it.

Richard opened the first meeting with a few words about the Greenhome project. "This is our first meeting since winning the competition," he said, "and I think it's worthwhile to ask ourselves at this point whether or not we now have to rethink the issues. Has anything changed? What is the exact thrust of this house from an architectural point of view?"

"Well," said Ed, "the focus is obviously environmental, but there are other considerations. I think the overall picture can be broken down into two subconcerns: Will it be energy-efficient, and can it be built and sold economically? Since the idea is to convince the building trades that energy efficiency is an achievable goal in house building, it seems to me that the major concept is that the house be reproducible."

"Not necessarily," John burst in. "This doesn't have to be a reproducible house; it's a prototype. We're building a house that people can go into and say, 'I want this component or that component.' They don't have to say they want to reproduce the whole house. On the other hand," he continued, "I think we should keep in mind that we have to use materials or products that are either on the market now or soon will be. In that sense, we have to be reproducible. We don't want to get into experimenting with oddball inventions. I had a call from a person who wanted to donate a gizmo that takes the greywater and recycles it back into the house. Now it's true that we want to use only half the water used in a conventional house, but we don't want to do it with a system that would reduce the overall attractiveness of the house."

"You'd have to go through a lengthy approval process for something like that," said Tony, patting the binder that contained the Code, "and I don't think we have the time. Besides, collecting greywater and making it drinkable is against society."

Tony's pronouncement and John's concern for the Greenhome's attractiveness reminded everyone that the house they were designing had to exist in the real world. More important, it brought up one of the most fundamental questions pertaining to the Greenhome, articulated almost immediately by Ed: "Just how far," he asked, "do we want to push people's perceptions about what a house can be?"

☐ ☐

The short answer to Ed's question is, Pretty far. The long answer is more difficult. For example, before it could be decided what kind of basement the Greenhome should have, the design team had to discuss whether the house ought to have a basement at all. From a salesperson's point of view, a house has to have a basement, period. From an environmental point of view, basements are energy disasters: in most conventional houses, up to 30 percent of the heat loss in winter is through the foundation, usually through poorly sealed gaps where the foundation wall and the house wall meet.

But even from a commonsense viewpoint, we should ask ourselves why Canadians are so addicted to basements. Rob Dumont, when he flew in from Saskatoon for a later Greenhome meeting in Waterloo's City Hall, became quite exercised over this addiction: "It's crazy," he said. "Why do we have to build our houses over holes in the ground? We dig a hole, the hole fills up with water, we pump the water out and build the house, and then we spend the rest of our lives pumping out more water. Why not just build on top of the ground in the first place?"

Rob is far from alone in his objection to a full basement. "A full basement," write John Cole and Charles Wing in their classic 1976 owner-builders' manual *From the Ground Up*, "is essentially a large and expensive concrete-lined well that we try to keep dry." Cole and Wing lament the fact that "9 people out of 10 consider a house without a full concrete basement as low-class construction suitable only for vacation homes or farm sheds." We could say, Who cares? But as often as not, all nine of those people work as mortgage officers at the local bank, and their attitude usually translates as no basement, no mortgage.

Before the age of superinsulation, full basements made more sense. They started out as root cellars under the kitchen, accessed by a trap-door, for storing root vegetables such as potatoes and turnips after the fall harvest. They were expanded to store fruit, either in bins or in Mason jars, and expanded again in northern climes to hold firewood and coal. When furnaces came along, the natural place to put the behemoths was in the basement, since warm air rises, which required further expansion and access through a separate and usually sunken side door. This introduced a fine irony: waste heat from the furnace effectively destroyed the basement as a root cellar. And filling it with wood and later coal and later still oil tanks made a basement a pretty dirty place, so it became a kind of necessary evil—good for storage and not much else.

But when we consider what most basements are used for today, Rob's argument makes a lot of sense. Let's take a walk down into my own basement. Watch your step: don't poke your eye out on that ski pole sticking out over the stairway, and here, let me move this chain saw out of your way. I meant to clean it yesterday. Okay, here's the laundry room, and a model of efficient use of space it is too. It's exactly 8 feet square—I didn't want to cut any more sheets of plywood or drywall than I had to; saved a lot of energy that way. Washer, dryer, laundry tub squeezed along one wall; ironing board and wooden clothes rack in the centre, fluorescent light on the ceiling (actually, on an exposed floor joist). We use the laundry room a lot, but there's no real need for it to be situated 8 feet below the ground, and when you think about it, we don't need a whole room set aside for it. The appliances could just as easily be arranged along any wall in the house. I remember a particularly sensible house I visited in Norway, in which the washer and dryer were in the bathroom, which was already plumbed; you couldn't do laundry while someone was actually using any of the other facilities (you had to walk through the shower to get to the washer and dryer), but that would be only an occasional inconvenience.

That large, rumbling thing in the centre of my basement is the furnace. Yes, it's an oil furnace — there's the 500-gallon tank over there. No, it isn't high-efficiency; even when it was installed, the manufacturer guaranteed only 65 percent, which means every time we pay for 500 gallons of oil, we actually get heat from 325 gallons. Watch your head, those ducts have sharp edges. They were put in after the house was built. Actually, there appear to be two or three generations of ducts down here; two of the cold-air returns don't seem to be hooked up to anything, and some of the hot-air ducts don't give off any hot air. Yes, it is quite warm down here. In fact, our basement is the warmest level of the house. That's because we haven't got around to putting that roll of insulation you see over there in the corner around the furnace and duct work. Of course, not all that heat is coming from the furnace; a good proportion of it is coming from the hot-water heater. It's an electric hot-water heater; people I know who work for Ontario Hydro call them cash cows.

The rest of the basement is a very useful storage area for some very useful things: old fold-out cots, stored here in case the Latvian army comes for a sudden visit; boxes of books for them to read when they get here; extra empty jars; my collection of broken chairs; stuff no one recognized the worth of at our last yard sale; other stuff we picked up for a song at our neighbours' last garage sale. Oh, yes, there's my workbench; here, I'll move this pile of old skates so you can get over to it. See how neatly it's covered with tools? And there are the shelves for cans of paint, surrounded by cans of paint. We desperately need a whole floor for this stuff, don't we? Just to keep it out of sight.

Houses without basements are simpler and therefore less expensive to build. Saskatchewan Conservation House, for example, was built on concrete pillars embedded in the ground to support "grade beams" —

reinforced concrete beams at ground level upon which a wooden floor was constructed. The floor was built over a polyethylene vapour barrier and insulated with 9½ inches of cellulose fibre for an R-value of 30. The mechanical room was upstairs against the north wall – solar heating does not involve Cadillac-sized furnaces and tanks big enough to fuel an icebreaker.

Another Saskatoon house, built in 1983 and described by Rob in *Watershed* magazine, had a preserved-wood foundation (which I'll discuss later in this chapter). The owner-builders of the house used a horizontal skirt of insulation to keep the frost away from the footings and tied the skirt into the insulated wall above, creating an enclosed crawl space that never went below freezing. In the old days, before adequate insulation was available, foundations had to go down to below the frost level to avoid spring shifting and cracking. In Saskatoon, the frost level was 8 feet down, so full basements made some sense. "With the new insulating materials," Rob wrote, "it is possible to limit the depth of frost penetration and thereby use a much less expensive surface foundation." The mechanicals went into a small storeroom on the main floor. "We missed the storage space initially," the owner told Rob, "but I'd say it's not a great loss, in that a basement would take as much effort to build as would building additional space above grade. Incidentally, the crawl space stays at a temperature of about 5 degrees C and serves as an excellent storage area for vegetables." Full circle.

Modern insulation materials and techniques have also made it possible to build a house directly onto a concrete slab poured on grade, as Greg Allen was doing on Amherst Island in the 1970s and as Nova Scotia architect Don Roscoe did in the mid-1980s; he perforated the concrete-slab foundation floors with ducts that fed a simplified heat-storage system. Without a lot of effective insulation under the slab, building a house on it would have been impossible – dampness and coolness would have made the house unlivable. But as rigid foam insulation that could bear the weight of 5 inches of concrete came on the market, slab-on-grade buildings began to make sense. Roscoe poured his slab over an inch of clear stone covered by 2 inches of SM extruded polystyrene and a 6-mil polyethylene vapour barrier. "You need the function a basement gives you," says Roscoe, referring to a basement's use in heating a house, not to its usefulness as a place for keeping old tennis balls, "but if you want to use the concrete in a foundation for heat storage, there's no point in having it in the basement when you live on the main floor. If you move the basement up and put it alongside the building, you can use it for a garage and a workshop and whatever else."

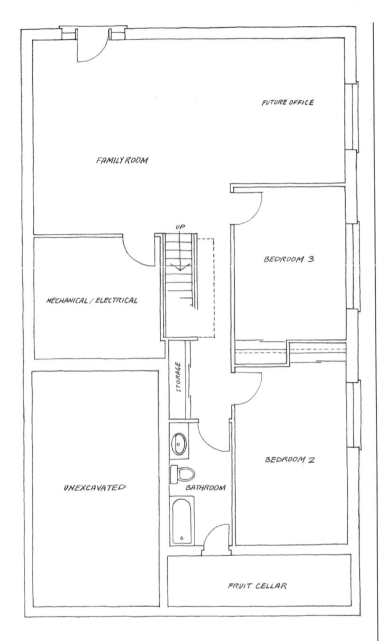

FUTURE OFFICE

FAMILY ROOM

UP

BEDROOM 3

MECHANICAL / ELECTRICAL

STORAGE

UNEXCAVATED

BATHROOM

BEDROOM 2

FRUIT CELLAR

The Greenhome's basement actually functions like most
houses' second floor, incorporating two bedrooms and an
office that are naturally warm in winter and cool in summer.
Three large windows on the above-grade south side provide
passive solar heat; the other walls are earth-bermed.

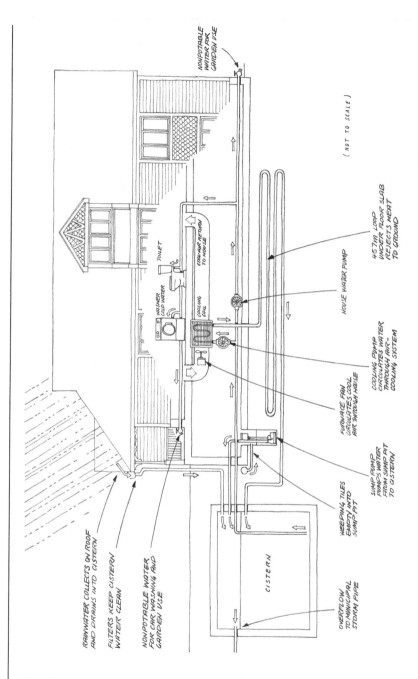

(NOT TO SCALE)

NONPOTABLE WATER FOR GARDEN USE

457m LOOP UNDER FLOOR SLAB REJECTS HEAT TO GROUND

COOL-AIR RETURN TO HOUSE

TOILET

WASHER COLD WATER

COOLING COIL

HOUSE WATER PUMP

COOLING PUMP CIRCULATES WATER THROUGH AIR-COOLING SYSTEM

FURNACE FAN CIRCULATES COOL AIR THROUGH HOUSE

SUMP PUMP PUMPS WATER FROM SUMP PIT TO CISTERN

WEEPING TILES EMPTY INTO SUMP PIT

RAINWATER COLLECTS ON ROOF AND DRAINS INTO CISTERN

FILTERS KEEP CISTERN WATER CLEAN

NONPOTABLE WATER FOR CAR WASHING AND GARDEN USE

CISTERN

OVERFLOW TO MUNICIPAL STORM PIPE

The Greenhome's low-tech water system includes a 4,000-litre cistern that fills with rainwater from the roof and supplies the house's toilets, clothes washer and garden hose.

We are, however, still at the perceived-value stage in house design, for the most part. We like the idea of empty space—we admire potential. To us, an attic is not an empty space in which we can store excess heat from the furnace until it can be safely vented outdoors; it is the study in which we are going to write our memoirs. Similarly, a basement is the future family room where we will all sit down together and watch archival footage of *The Ed Sullivan Show* or snuggle in the glow of a crackling recycled-newspaper fire. The Greenhome committee decided that the Greenhome would have a basement, but it would be a basement with a difference. First, it was to be only half a basement, and second, it was to be fully integrated with the rest of the house.

The main problem with basements is dampness. My basement, even though it looks dry, smells damp. It has no drainage; at the time it was built, the customary practice was to run the downspouts from the eaves-troughs through the foundation walls into the basement, to feed directly into a pipe leading to the city's storm sewer. That sensible practice eliminated the need for weeping tiles in some builder's mind. However, that sensible practice was outlawed a few decades ago by the city, so the downspouts now curve gracefully outward to discharge their water about 18 inches from the foundation wall. Malcolm Wells, in his 1981 book *Gentle Architecture*—a plea for the acceptance of green house design—estimates that a 26-acre shopping plaza discharges 600,000 gallons of water into its storm sewer for every inch of rain that falls on it; my calculator tells me that my 60-square-yard roof therefore deposits about 286 gallons of water beside my basement wall every time it rains an inch, and most of it sinks into the ground beside my basement wall.

In the Greenhome, at least some of it will go into a cistern. Cisterns have gone out of fashion in the past few decades, but in early Canadian farmhouses, they were important features. Basically, a cistern is a holding tank for rainwater; built into the corner of a basement and made of concrete or sometimes wood parged with pine tar or just tar, cisterns were filled with water from the roof, and the water was used for just about everything water is normally used for in a house except drinking and cooking. Laundry water came from the cistern, as did bathwater, water for the garden and, later, water for flushing the toilet.

A farmhouse I lived in during the late 1950s, in Dalston, Ontario, had a cistern in the root cellar, directly below the kitchen, and a small hand pump on the counter beside the kitchen sink. We had a well for drinking water, but no hot-water heater at all except for a small reservoir in the woodstove, and I can still smell the sweetly rank odour of the water we pumped up for washing our faces in the morning. It was, in retrospect, a perfectly adequate system. Years later, when I lived in a country house north of Kingston, Ontario, my neighbour, who at the age of 86 was only slightly older than his house, kept a cistern in the basement filled during dry spells and in winter by pumping water up from a nearby lake. Well water supplied the cold taps in his kitchen and bathroom; everything else came from the cistern, including hot water and flush water, which had its own plumbing hooked up to a separate pump.

The Greenhome's cistern will be a modern version of its 19th-century predecessors. It won't be in the basement but will be buried under the front lawn, although its purpose is still to provide free water. Airtight houses ought to have space-cooling as well as space-heating systems; in the Greenhome, water from the cistern will circulate through 1,500 feet of plastic pipes laid under the basement floor, where it will be cooled to about 55 degrees F. It will then be pumped up into a series of copper coils running through the house's heating ducts, where it will cool the air blown through the ducts by the furnace fan.

Water from the cistern will also feed part of the house's greywater system: pipelines leading from the cistern to the garage and backyard will be used to water the garden and wash the car, another pipeline will be used to flush the toilet, and yet another will supply cold water to the clothes washer. To do this, John Kokko calculates he'll need a 4,000-gallon tank with an overflow gravity feed to the storm drain for wet seasons and a float inside connected to city water to refill the tank when its level sinks below a certain point.

Placing the cistern outside the house was more complicated than you'd think. It had to go outside the house, because if it were inside, it would be heated, and heating a concrete tank holding 4,000 gallons of rainwater to room temperature is not an efficient use of natural gas, especially when you then turn around and cool it again in order to cool the house. Burying it below the frost line so that its water won't freeze in the winter disturbs the environment with one more hole, which is not a good idea either, and also means that the cistern has to be made out of concrete. Plastic cisterns are available, but they have to be filled with water before they can be buried so that the weight of the earth doesn't crush them. And it just doesn't make sense to fill a cistern with tap water, since its main purpose is to make use of rainwater. John checked the embodied-energy figures for concrete versus plastic, though, and found that they came out almost even. The design team opted for concrete.

Getting back to the question of the basement: I've said that the decision was to make the Greenhome's energy-efficient and useful. This was to be accomplished by using the most environmentally responsible basement-wall material available, by raising the south-facing basement wall half out of the ground so that the space could be warmed (and dried) by solar heat energy, and by situating as much living space in the basement as possible. This meant, essentially, making a traditional bungalow (one storey plus basement) into something more closely resembling a two-storey house. As it happened, Steve had just bought a house in the village of Conestogo, just outside Waterloo, in which the bedrooms were located in the basement. This turns traditional house layouts on their heads, but the arrangement made a lot of sense to Steve.

"When I thought about it," he says, "I realized that we've been doing it wrong all along. We build a two-storey house and put the bedrooms on the second floor, which is cool in winter and hot in summer, which means we have to spend energy heating them in the winter and cooling them in the summer. Basements are like that naturally. In my new house, the bedrooms in the basement are cool in summer and warm in winter." Putting the bedrooms in the basement also freed up a lot of space upstairs on the main floor for living and dining areas, a larger bathroom, a larger master bedroom with a walk-in closet and other cosmetically attractive features.

□ □ □

Having decided on a basement, the designers still had to wrestle with the question of materials: What is the most energy-efficient foundation-wall system available? The most common material used by professional builders in North America today is poured-in-place concrete. The builder levels the bottom of his excavation hole and makes shallow 2-by-6 forms where the outside walls will go, plus a few inside the floor space where the load-bearing walls will be, then calls in a cement truck to pour the footings. These are the solid platforms upon which the basement walls will rest. When the footings set, the builder builds another, more massive set of wooden forms, usually out of plywood and 2-by-4s, so the space between the forms is 8 or 10 inches thick and as high as the basement wall will be, about 8 feet. This space is first punctuated with rebar, vertical lengths of steel rod set about 15 inches apart, around which the cement will be poured. Concrete has plenty of horizontal, or compressive, strength—you could build a five-storey building on 10-inch concrete foundation walls—and the rebar provides lateral strength so that the walls don't get pushed inward by the force of the earth against their sides.

Builders and architects like poured concrete better than concrete blocks for a variety of reasons. Probably chief among them is that they just like the idea of poured concrete, so much so that in 1978, American builders were using 450 million cubic yards of it every year, for everything from highway overpasses to kitchen sinks. Here is Malcolm Wells, the guru of gentle architecture, rhapsodizing about poured concrete: "Imagine," he says, "having stone you can pour! It must be one of the most useful materials known to man. Used for the construction of roads, walks, plazas, dams, foundations and buildings, it may at some future time become the principal material by which archaeologists are able to recognize our era."

On a more practical level, builders like reinforced poured concrete for its lateral strength: no building blocks held together by mortar applied by mortal hands can push back against the earth as well as a poured-concrete wall braced with steel bars. Builders also appreciate its convenience: it takes a couple of hours for an experienced work crew to set up the forms, then they just call in the cement truck and go home for a day while the wall sets and cures. It is, as builders say, liquid rock. Cement is a mixture of clay and powdered limestone, after all, and making concrete from it is simply a matter of returning it to its original state; when properly combined with water, sand and gravel, it regains half its final hardness in three or four days and can be built on in two.

A third consideration, if you still need one, is that concrete blocks are hollow and difficult to insulate adequately for a northern climate. A 10-inch concrete block with nothing stuffed into the hollows has an R-value of 1.08. Defenders of concrete blocks tend to live south of the snowbelt. Kirchner Industries of St. Louis, Missouri, for instance, built a passive solar house entirely out of 8-inch blocks fitted with polystyrene inserts, called Korfil, and pebbled vermiculite, for a total R-value of 7.2. In St. Louis, that might make for a cosy house, but don't meet me in Winnipeg. Now it's true that a block wall can be insulated on the outside and the inside, but so can a poured wall. All things considered, the only real advantage blocks have over poured concrete north of the Mason-Dixon line is that one person can put them up. That is a boon for the solitary owner-builder, but it didn't carry a lot of weight with the Greenhome team. And as we shall see, concrete blocks are not the only foundation system with that advantage.

It should be mentioned that one of the real problems with concrete blocks—the weakness of the mortar that holds them together—has been addressed by the United States Department of Agriculture with a system called "surface bonding." With this technique, the blocks are laid in the normal way but without mortar (except for the first course, which is mortared to the footings); when the blocks are in place, both sides are parged with a mixture of cement and fibreglass strands, and that's it. Wall done. The exterior can be stuccoed, and the inside can be covered with drywall, but both are optional. The resulting wall has six times the lateral strength of a traditional concrete-block wall, it can be put up quickly, and the parging reduces the air penetration so well — from 496 cubic feet of air per minute for an unparged wall to 0.1 cubic foot after parging — that no interior air/vapour barrier is required. This is a fairly common practice now, but for the Greenhome, it didn't outweigh the other problems associated with concrete blocks.

At the design meeting, Werner and Ian had naturally assumed that the Greenhome's foundation would be poured concrete, and they were taken aback when John voted against the concept.

"What's wrong with concrete?" asked Werner.

"Concrete," John replied, "has 10 times the energy input that wood has." The energy embodied in concrete is enormous: it is, after all, a mixture of crushed limestone and crushed stone, and the heavy, diesel-burning equipment used to do all that crushing accounts for untold barrels of fossil fuel a year. Concrete is also very heavy and therefore requires a lot of energy to transport, both from the manufacturer to the supplier and from the supplier to the job site. This is reflected in its cost: $100 a cubic yard. And in water-conscious southwestern Ontario, the fact that a cubic yard of pourable concrete requires many gallons of water is another significant factor in the embodied-energy equation. This is true not only in southwestern Ontario, by the way; according to Rob Dumont, in Saskatchewan, the price of water is rising more rapidly than the price of energy.

The foundation system chosen by the Hamilton Advanced House, which is called the Neat Home, addresses the insulation problem in a unique way: they built forms for poured concrete, but before pouring the concrete, they suspended a layer of 2-inch Styrofoam SM and a layer of Baseclad between the forms and then poured the concrete on either side to make a kind of insulation sandwich. But the Greenhome team was opposed to using poured concrete.

The discussion opened into alternatives to poured concrete. No one mentioned concrete blocks. "Well," said Elizabeth dubiously, "there's preserved wood."

This was met with some vigorous head shaking around the table. Preserved-wood foundations (PWFs), as their name implies, are foundations made out of wood that has been treated with a preservative to prevent the wood from rotting. The first house to be constructed on a foundation made from preserved wood was at the Forest Products Laboratory in Madison, Wisconsin, in 1937, so it's not exactly a new technique. But it's a good one: the Madison house's foundations were checked in 1966, and according to the lab report, "after 30 years of use, this wood foundation was giving excellent service and effectively meeting its requirements." Admittedly, the requirements in 1937 were not rigorous. The wood used—it was a 3-foot wall made of 6-by-6 Douglas fir and 2-by-8 southern pine planks—was treated with creosote, a coal-tar product that has since been banned because it is highly toxic and leaches out easily into the surrounding earth. Think about all those treated railway ties that are used to make garden retaining walls.

In Canada, the first preserved-wood foundation was built by the NRC in Ottawa in 1960; two more houses were built there, the Mark III in 1961 and the Mark IV in 1964. Mark III used creosote-treated lumber and plywood; Mark IV used wood treated with a then-new preservative called pentachlorophenol, which turned out to be even more toxic than creosote.

"I'm still worried about the toxic content of PWFs," said Elizabeth. The preservative now used to make PWFs is called CCA, which stands for chromated copper arsenate, a mixture of the oxides of chromium, copper and arsenic. PWF manufacturers claim that CCA does not leach out of treated wood into the ground, but there remains the question of how much arsenic oxide escapes into the air—and workers' lungs—during the manufacturing process itself. And what happens to the dust from a CCA-treated 2-by-4 when a carpenter cuts it with a circular saw?

Richard was concerned about the innovativeness of PWF. "Is preserved wood like an Edsel?" he asked at the meeting, bringing the discussion back to the Greenhome's role as a demonstration house, "or is it a good idea that should have more public acceptance? From Elizabeth, I gather there's a question about toxicity, and from you, Ian, I'm hearing that a wood foundation is such a weird idea that builders will reject it."

"Well," said Ian, "you can poke holes in wood from a number of perspectives. For instance, there's the life-span issue. Last summer, I helped my father build a cottage up in Kincardine, and we used PWF. It was perfectly suited to that area, which is all dry sand. But down here, where the ground is almost always wet, I don't think you're going to have a structure there 50 years from now if you use wood. I don't see the advantage of it. If it's a question of construction time, I can have a team come in at 2:30 to put up the forms and we'd be ready to pour at 3:40."

The first house in Canada with a CCA-treated foundation was built by Beaver Lumber in Morinville, Alberta, in 1967, and since then, there have been more than 100,000 PWF houses built in Canada alone, mostly in the Prairie Provinces, where the soil is naturally sandy and dry and aggregate for making concrete is in short supply. The Prince Edward Island Advanced House is built on a PWF system modified by Ottawa engineer Doug Walkinshaw. Called the Echo system, it deals with the problem of dampness by incorporating a mechanical ventilation system that exhausts the moisture and gas that build up between the exterior and interior walls.

In central Canada, however, where you can have a cement truck at your work site within an hour and where the ground is often clay-based and the water table high, a PWF is still something of a novelty, and builders tend to be skeptical of novelties. Werner told me one day that if the design committee had decided on a preserved-wood foundation, he would have backed out of the project, and other builders I've talked to feel much the same way. "Pressure-treated wood rated for ground contact is okay," writes Jim Locke snidely in his book *The Well-Built House*, "unless the structure above it is quite expensive or is an integral part of the house." In other words, if your building is going to house anything heavier than a chicken, "you'll want the security of concrete." Jim Locke is the builder of the house that Tracy Kidder wrote *House* about.

To be fair, there is some basis for Ian's concern about a PWF's lasting power. Preserved wood is made by impregnating ordinary lumber with fungicide in a kiln in which the wood is heated to 100 degrees C at 10 times atmospheric pressure – which is why it is sometimes called pressure-treated wood. Fungus is the chief and almost sole cause of wood rot. Wood cells contain cellulose and lignin, two favourite foods of a surprising number of fungi. A fungus needs both water and air to grow – wood kept completely underwater will last for centuries but will rot within weeks of being raised to the surface. Rather than protecting wood from water and air, which is what creosote did, CCA just kills the fungi.

But there are problems with pressure-treated wood that could affect its longevity. One is quality control: the Canadian Standards Association (CSA) set fairly rigid standards for PWF, especially for the part of it that will end up below grade. According to the CSA, below-grade wood must contain preservative penetrating to at least ⅜ inch from each surface, and each cubic metre of wood must have soaked up at least 6.4 kilograms of CCA. Then a consortium of PWF manufacturers came along and imposed a standard on themselves that was, guess what, less rigid than the CSA's: they said ³⁄₁₆ inch was enough penetration for them. This is called the PS-1 rating. Now, PS-1 is all right for above-grade work – porches, decks, fences resting on concrete posts, et cetera – but subgrade lumber must meet CSA standards and must be stamped

PS-1/PWF if it is to have a hope of lasting the 40 years the manufacturers claim for it (or of passing a building inspection).

There are, fortunately, a number of alternatives to poured concrete besides PWF, some of which are variations on concrete blocks. One is a Canadian product known as Sparfil. In 1974, Jeff Sparling, a construction engineer living in Cobourg, Ontario, set out to find another use for the tiny beads of expanded polystyrene that are normally pressed into sheets and used as insulation – the same stuff those crumbly coffee cups are made of. Sparling mixed the beads with ordinary cement, at a ratio of about 60 percent beads to 40 percent cement, and poured the result into moulds to produce large rectangular panels that he felt would be very useful in commercial construction. The panels had an R-value of about 1 per inch, which was not impressive to the low-energy people, and they never really caught on with the commercial builders either. Two years later, he experimented with shaping the bead/cement mixture into the form of standard concrete blocks, then he filled the hollows in the blocks with Styrofoam inserts, and voilà, the Sparfil block.

Sparfil blocks did catch on with owner-builders: they were light and could be cut with a carpenter's saw, and surface bonding eliminated the need for mortar and therefore for a bricklayer. With their R-value of 18.5 for a 10-inch filled block, low-energy builders liked them too.

But Sparfil blocks are made with expanded polystyrene, which is blown with pentane, a fact that eliminated Sparfil from serious consideration for the Greenhome. A variant of Sparfil, called Durisol Wall-Forms, avoided the CFC problem: Durisol blocks are similar to Sparfil but are made with a combination of cement and wood shavings. The blocks are dry-stacked, and instead of expanded polystyrene, the hollows are filled with poured-in-place concrete. The resulting wall has the structural integrity of concrete and the insulative value of wood shavings, and its manufacture does not destroy the ozone layer. But Durisol blocks have been around since 1950, so the innovation factor worked against them. If they haven't sparked a revolution in the building trade by now, they probably won't.

A more recent version of Durisol is a system called Argisol. Actually, Argisol is a kind of hybrid between Durisol and Sparfil: it consists of foam blocks filled with poured concrete. I was sitting in the back of the framer's truck one day to get out of the rain when two fellows dropped by to ask about the Greenhome. They introduced themselves as Alberto and José Menendez; they were building a small triplex over by the University of Waterloo and were using Argisol blocks for the foundation. I went by the next day to have a look.

Known as "Argisol, the energy-conscious system," it was developed in Switzerland in the late 1970s – as Alberto said, "they ran out of natural building materials in Europe 20 years ago, so they had to come up with all this new stuff." It has an R-value of 22, which is impressive, and uses a flame-retardant expanded polystyrene that is not, according to the brochure, CFC- or HCFC-blown. Each block is 10 inches wide, 10 inches high and 40 inches long and consists of two panels connected together with metal webs. When filled with poured concrete and rebar, the blocks form an insulated concrete wall that is parged and moisture-proofed like any other system.

Argisol has a lot of pluses for the owner-builder of a house with no basement and easy access to a cement truck with a pump. It is quick and easy to install – Alberto and José, with help from an Argisol representative, put up 180 feet of 4-foot foundation wall in four hours, including the time it took to pour the concrete. The blocks are extremely non-porous: the Code specifies a maximum water absorption by volume of 4 percent; Argisol came in at 1.1 percent. And the system is not prohibitively expensive: José says they spent $3,300 for the whole foundation, including $1,000 for 10 yards of concrete.

But if the basement wall is to be full-storey height, as it is in the Greenhome, then it has to be braced with wood shoring or with the special bracing members manufactured for the system. This means putting up almost as many forms as you would for poured concrete, not much of an improvement in technique over traditional construction. And for the Greenhome, another disadvantage was the foam, a petroleum by-product even if it wasn't produced using CFCs or HCFCs.

One final system was discussed that seemed at first to satisfy even the Greenhome's criteria – it's low-energy, it uses recycled or waste materials, it's Canadian, and it's brand-new. It's called cementatious wood and is a product of C-MAX Technologies Inc. of Vancouver. Like the old Durisol blocks, cementatious wood is made of a mixture of cement and wood fibres. By varying the kind and ratio of mineral and plant fibre, however, C-MAX was able to come up with a variety of building materials, which can be used almost exactly like dimensional lumber (that is, they can be sawn and nailed). Instead of being shaped into blocks, the different formulations can be made into boards, beams, panels, and so on; in other words, a foundation of cementatious wood could be made exactly like one of preserved wood.

One advantage of cementatious wood is its low cost, which is the result of being made of recycled materials. "Present-day harvesting and manufacturing processes," says C-MAX in its literature, "reject 40 to 50 percent of wood or plant species either in whole or in part as nonusable.

In addition, there presently exists an ever-increasing volume of non-recyclable wastepaper worldwide. . . . All these fibrous materials are inexpensive and, when processed into various particle sizes, can make up 60 to 90 percent by volume of each of the C-MAX formulations, thereby assuring cost effectiveness in the end products." The low cost should be attractive to builders; the use of recycled materials was attractive to the Greenhome design team.

C-MAX Technologies produced a kind of cementatious-wood kit — pulverized magnesia, dolomite and plant fibres in one container, liquid ammonium polyphosphate in another — and was hoping to simply ship the raw materials to manufacturers who would then combine them, shape them into whatever form a customer ordered, deliver the product to the customer, collect the money and send part of it back to C-MAX. The problem was, no manufacturers were lining up to get on board, so the product was unavailable in the east. Shipping it from Vancouver to Waterloo would have pushed its embodied-energy factor too high, so despite John's interest in using cementatious wood in the Greenhome, he had to scrap the idea and look around for something else.

□ □ □ □

I HAVE NEVER FULLY UNDERSTOOD HOW A FOUNDATION HOLDS UP A HOUSE.
— JOHN N. COLE AND CHARLES WING, *FROM THE GROUND UP*, 1976

The basement the designers finally chose arrived at the building site on October 5 on the back of a 45-foot flatbed truck, in the form of 16 precast concrete panels. Each large panel was 8 feet high and 16 feet long and weighed 6,400 pounds. The smaller panels were half that size. Behind the truck came a crane for getting the panels off the truck and down into the excavation. Jerry, the crane operator, positioned his machine at the southwest corner of the hole, and we were ready to begin.

The panels were made by Lake Huron Precast, a small company based in Zurich, Ontario, owned and operated by Bert van Kruistum and his cousin, Peter Damsma. Peter had been a framer until the bottom fell out of the custom housing market in 1990 and he threw his lot in with Bert. "There was no money in it for a good framer," he said as he checked the footings at the corners of the foundation. "Lots of people got into the business when the economy went down — anyone with a circular saw and a square was calling himself a framer. And most of them could get

Concrete requires more energy to produce than any other building material, so the Greenhome used precast panels containing half the concrete of conventional poured-in-place foundations. Each panel was 8 by 16 feet, weighed 3 tons and had to be craned into position.

away with it with tract houses. Those things are just stamped out. Put one of those guys on a roof with 17 valleys, and they wouldn't know where to begin. But no one's building custom houses anymore."

Bert and Peter hope that their system will be attractive to tract-house builders. They have used it in commercial buildings, but the Greenhome is the first time it's been adapted to residential construction. Their foundation is made by bolting precast panels together instead of pouring the concrete in place. It has many advantages over poured foundations: the panels are cast at a factory in adjustable metal forms that can be used indefinitely and changed to fit any builder's specifications; it doesn't require continuous footings but rather a series of concrete pads, one pad under each end of the bolted panels; the panels are smooth on the outside and waffled on the inside, with the result that they use 50 percent less concrete—a compelling boon for energy-conscious builders—without giving up any compressive or lateral strength, and the cavities can be filled with insulation for added R-value.

In fact, they are stronger than poured-in-place concrete: the Building Code requires that concrete in poured foundations be at least 20 megapascals (MPa), or strong enough to resist 4,000 pounds of pressure per square inch; Bert and Peter mix their concrete to 35 MPa (7,000 psi) by using less water in the mixture, a chemical superplasticizer and a higher cement-to-aggregate ratio. "This gives it what we call high-density strength," explained Bert. "At 20 MPa, concrete acquires 50 percent of its strength in four or five days. At 35 MPa, that time is reduced to 8 hours. We pour the concrete into the moulds at 2 o'clock in the afternoon, and we can lift them out first thing the next morning."

What this means to builders is that they can order their foundations one day and have them delivered to the building site the next afternoon. This is almost as fast as building forms and pouring concrete and requires fewer workers. With a heavy crane and proper footings, Bert and Peter can install a foundation in four or five hours. The system has the strength of a full concrete wall but uses only half the concrete, and it impressed the design team.

The first time I saw one of the panels was in the back storeroom at Enermodal—John had procured a sample from Lake Huron Precast. Standing in the storeroom among pieces of ventilation ductwork and heat-recovery-unit housings, it looked like a piece of Stonehenge rising over the scene of a plane crash. This was a cutaway sample, though, and consisted of a 2-foot slab standing about 5 feet off the floor. At the cutaway, I could see that the concrete was very heavy and solid, not porous like a sidewalk, and speckled with large pieces of gravel. It was U-shaped, with the full 8-inch side pieces, each 3 inches wide, and the 18-inch centre reduced to a thickness of only 2 inches. At the cutaway top, I could also see the ends of rebar—two rods in each vertical column—and there was also a rebar web in the thin centre section.

"That's how they reduce the amount of concrete," John had said, pointing to the recessed centre. "Each full panel consists of four of these sections, so a 16-footer uses only about 1¼ cubic metres of concrete, or half the amount of a full 8-inch poured foundation wall. And even factoring in the steel rebar, it has less embodied energy than poured. And the big plus for us," he said, indicating the bits of gravel visible at the cutaway, "is that this aggregate is recycled. Usually, when a cement truck gets back to the yard at the end of the day, it has a bit of extra concrete in it that they just dumped out into a heap and left there. With the economy the way it is, the companies are realizing that that heap of concrete represents lost profits, so now the drivers are pouring it out into a 4-inch slab, and when it dries, they crush it up and sell it as aggregate. Lake Huron Precast is using this recycled aggregate to make these panels, and we're using it in our weeping-tile bed. It's perfectly good stuff that used to be just waste."

A few days before Bert and Peter arrived with the panels, the footings were poured and the weeping tile was laid. A weeping-tile bed is a kind of underground moat that surrounds the foundation at the footing level, the purpose of which is to drain water away from the house. The danger of water around a basement is twofold. Obviously, a lot of water sitting on the outside of a basement wall is not good, especially if the moisture barrier is not leakproof (and 40 percent of them are not). But even if the moisture barrier works, having water-saturated ground outside the wall is dangerous, because it will freeze during the winter and expand, pushing against the wall with a great deal of force. Worse than that, if you can imagine anything worse than that, when the ground thaws in the spring, it will shrink, leaving an open space between the ground and the wall into which meltwater and rainwater will run. The water will then go right down to the footings, softening the ground *under* the footings and causing the whole house to sink. So all in all, weeping tiles are a good thing.

In this case, the weeping bed is a layer of recycled concrete aggregate, lining a trench outside the foundation footings. On this layer is laid a 4-inch weeping tile or plastic drainage pipe with perforations at the top so that water can seep (or weep) into the pipe and be carried off to the storm drain. The pipe is then buried under a thicker layer of finer crushed stone or aggregate, which acts as a filter so that only water, and not silt, reaches the tile. If silt is allowed to reach the tile, it will clog up the perforations and keep the water out of it. The Greenhome tile was actually wrapped in a fine mesh sock, like cheesecloth, to further protect the inside against silt; another layer of cheesecloth was placed over the gravel as a third precaution. Then the whole thing was ready for backfilling as soon as the wall was in place on the footings.

The footings were really just 3-foot-square concrete pads poured into 6-by-6 forms at each corner of the building and at 16-foot intervals along the walls, wherever two panels would meet. When the large 35-ton crane arrived at 1 p.m., the panels were hoisted from the flatbed truck on pins fitted into holes left at the top of each panel. The crane hovered the panels a fraction of an inch over the footings while Bert and Peter heaved and cajoled them into position, using crowbars, bits of lumber, their feet or anything else that was convenient. When a panel was perfectly lined up with the chalk lines Bert had placed on the pads, the crane lowered it fully and the pins were removed. Now the panel was bolted to the one next to it through four holes left in the sides, using 12-inch bolts and an ordinary ratchet wrench. Between each panel, Peter inserted a strip of Waterstop an inch in from the outside surface of the wall to seal the joints between panels and keep the wall watertight.

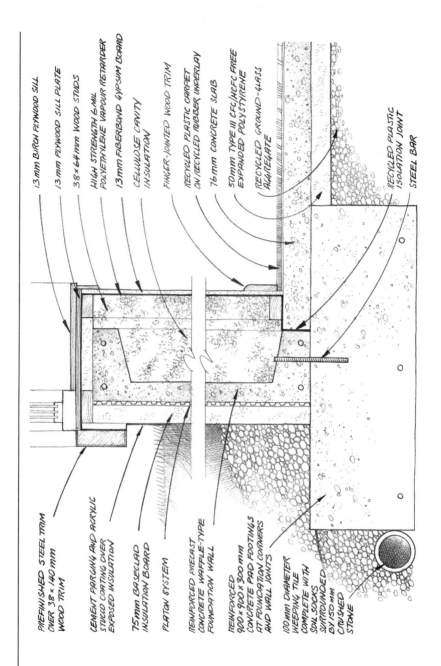

13mm BIRCH PLYWOOD SILL

13mm PLYWOOD SILL PLATE

38×64mm WOOD STUDS

HIGH STRENGTH 6 MIL POLYETHYLENE VAPOUR RETARDER

13mm FIBERBAND GYPSUM BOARD

CELLULOSE CAVITY INSULATION

FINGER-JOINTED WOOD TRIM

RECYCLED PLASTIC CARPET ON RECYCLED RUBBER UNDERLAY

76mm CONCRETE SLAB

50mm TYPE III CFC/HCFC FREE EXPANDED POLYSTYRENE

RECYCLED GROUND-GLASS AGGREGATE

RECYCLED PLASTIC ISOLATION JOINT

STEEL BAR

PREFINISHED STEEL TRIM OVER 38 × 140 mm WOOD TRIM

CEMENT PARGING AND ACRYLIC STUCCO COATING OVER EXPOSED INSULATION

75mm BASECLAD INSULATION BOARD

PLATON SYSTEM

REINFORCED PRECAST CONCRETE WAFFLE-TYPE FOUNDATION WALL

REINFORCED 900 × 900 × 300 mm CONCRETE PAD FOOTINGS AT FOUNDATION CORNERS AND WALL JOINTS

100mm DIAMETER WEEPING TILE COMPLETE WITH SOIL SOCKS SURROUNDED BY 150 mm CRUSHED STONE

The foundation panels were set on 3-by-3-foot concrete corner pads. Their hollow centres were filled with blown-in insulation made of recycled newsprint. The basement floor pad was poured over non-CFC-blown polystyrene sheets.

As he worked, Bert pointed out another benefit of using precast panels. "The Code calls for poured-in-place concrete to be 20 MPa, as I said. But concrete doesn't become water-resistant until 25 MPa, which is why you also have to paint it with tar on the outside. These panels, at 35 MPa, are already very water-resistant. You don't really need a moisture barrier on them at all."

But the Code calls for a moisture barrier, so the Greenhome has to have one. Tar (actually a liquid form of asphalt, but everyone calls it tar) was out, of course, being a petroleum product. The search for an alternative turned up two possibilities. One was a thick rubberlike sheeting called Mel-Rol, which, according to the specifications that came with the sample, consists of a layer of bitumen on one side and a plastic covering on the other. Bitumen, however, is defined in Webster's dictionary as "a mineral substance of a resinous nature and highly inflammable, appearing in a variety of forms which are known by different names." Among those names are naphtha, petroleum tar and asphalt.

In the end, the designers settled on a product developed in Norway called the Platon damp-proofing membrane. Platon is a black, high-density polyethylene, four times thicker than the 6-mil poly used for vapour barriers, pressed with a pattern of rounded dimples that keep the membrane a few millimetres away from the wall. These dimples, says Platon's data sheet, solve "the age-old construction problem of the waterproof moisture membrane acting as a cold-side vapour barrier."

Translated out of engineerese, this means that the Platon system avoids the problem inherent in moisture membranes like tar, which adhere directly to the outside of the concrete wall that has a stick-frame insulated interior wall attached to it (as most habitable basements do). Paint the outside of the wall with tar, and the tar keeps rainwater from wicking through the concrete into your house. That's good. But it also prevents moisture which accumulates in the wall from the heated interior or which comes up through the footing from wicking through to the outside. That's bad, because it means that condensation inside the wall stays there, soaking the insulation and eventually rotting the wood.

The dimples on the Platon system prevent moisture from being trapped inside the wall by leaving an airspace between the membrane and the outside of the wall, down which moisture can run and be directed harmlessly into the weeping bed. There are other advantages as well, according to Roy Frater, who installed the membrane on the Greenhome a few days after the concrete panels were up. Frater, a technician with Engineered Basement Solutions, believes that the Platon system is the best thing to hit house building in 20 years.

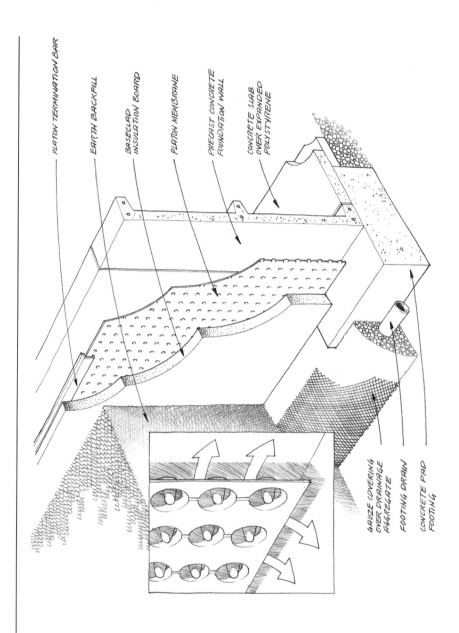

PLATON TERMINATION BAR

EARTH BACKFILL

BASECLAD INSULATION BOARD

PLATON MEMBRANE

PRECAST CONCRETE FOUNDATION WALL

CONCRETE SLAB OVER EXPANDED POLYSTYRENE

GAUZE COVERING OVER DRAINAGE AGGREGATE

FOOTING DRAIN

CONCRETE PAD FOOTING

Conventional foundations are coated with tar to prevent
moisture from leaving the wall. The Greenhome's foundation
vapour barrier, the Platon system, provides an opportunity
to avoid using an environmentally costly fossil fuel. It consists
of a dimpled plastic membrane that allows moisture to wick
out of the wall and drain down to the weeping tiles.

88

"For one thing, it covers up so many defects in the foundation wall," he said, "cracks and holes and such that have to be filled before being painted with tar. This membrane is so thick that there is no give when the hole is backfilled. And apart from allowing the wall to breathe moisture, it also allows trapped gases to escape—methane is a big concern here in the Waterloo area, and radon is a serious problem in other parts of the country."

Radon is a serious problem in a great many parts of North America. It is an odourless, colourless natural by-product of radioactive decay in the Earth's bedrock that seeps into basements and stays there. As radon decays (its half-life is 1,602 years, so don't wait for it to go away by itself), its daughters attach themselves to dust particles and are inhaled. Two of these daughters emit alpha particles that are up to 20 times more dangerous than X-rays. The danger is lung cancer—estimates in Canada are that radon may be the cause of between 500 and 2,000 cancer deaths a year, and in the United States, the Environmental Protection Agency (EPA) believes that radon is the second biggest cause of lung cancer after cigarette smoking.

The best defence against radon is good ventilation—houses equipped with heat-recovery ventilators (as all superinsulated houses will be when the 1995 Code comes out) are relatively safe. But since concrete is made with rock aggregate, radon can actually be emitted from the foundation walls themselves. A good interior vapour barrier can keep most of it outside the living area, but many basement walls are unfinished, and many finished basements have vapour barriers that have been punctured with everything from drywall screws to utility outlets. "With the Platon system," says Frater, "you can tap into the airspace between the plastic and the concrete and vent the radon out."

Frater added that this system has been used in Norway for the past 25 years and is now required by the Norwegian building code. "You can't use tar there at all," he said. "I foresee this or something like it being standard here in the next 5 to 10 years."

The Platon system is also easy to install. It comes in rolls that are 2 metres wide and 20 metres long. The wall is first prepared by the application of a "flood boot" along the bottom where it hits the footings—a strip of thinner high-density poly that is bonded to the wall with a compound called Platonite and curved out over the footings to ensure that water running down the wall under the Platon membrane is directed out into the weeping bed and not under the floor slab. After the flood boot is applied, the membrane is simply rolled around the house and tacked loosely into position with concrete nails encased in plastic seal

89

plugs ("Don't put it on fiddle-string tight," advised Frater). Then a strip of galvanized steel called the termination bar is nailed into place along the top edge of the membrane to keep rain and snow from running down into the airspace. Where there are joins in the membrane, the dimples lock into each other at the overlap, and no caulking is needed. "Three people can do a whole house in under three hours," said Frater.

At the Greenhome, two of Frater's workers wrapped up the foundation wall in half a day. When they were finished, the Greenhome's basement was ready to receive its upper deck. Within the basement's walls, in its subfloor, 6 inches of sand covered 1,500 feet of geothermal tubing for circulating the water from the cistern. Above the sand are a few inches of aggregate for drainage in case the basement ever fills up with water. Instead of gravel, the aggregate would consist of a few truckloads of crushed green glass from a local recycling company. One of the main setbacks of glass recycling has been the lack of a use for coloured glass. Mountains of it have been stockpiled from millions of Blue Boxes nationwide. Crushing it up and using it as aggregate in places where no one will ever have to walk on it—under concrete floor slabs, for example—is an innovative use for a waste product that perfectly fits the Greenhome matrix. Unfortunately, the first load, which came from a local recycler, was so full of tin cans and plastic bottles that it looked as though it had been dumped unsorted from a curbside recycling truck. And the glass was not so much crushed as just broken: big, jagged pieces glinted hungrily in the early October sunlight, waiting for someone without proper work boots to step on them.

"Doesn't look like good drainage material to me," said Peter.

"Looks like a downright work hazard to me," said Bert.

But by the end of the first week in October, John had found a better supplier, and the excavation floor had been covered and raked, ready for the concrete slab. Outside, where the wind had once again shifted to the north, the air was cooling even though the sun shone brightly. But as I stood in the shelter of the open basement while the crane operator raised his hydraulic stabilizers and prepared to go home, it was still warm and comforting. With the plans in my hand, I could stand at the centre of the basement and see where the fruit cellar was going to be (at the front of the house under the porch) and where the mechanical room was located (over there behind the garage). There was the walkout from the downstairs living room, and there was the concrete footing for the central interior load-bearing wall, at the spot where the stairs would end. Even with the clear blue sky over my head, the Greenhome was beginning to look like a house.

FRAME WORK

□

THE BUNGALOW IS THE RENEWAL IN ARTISTIC FORM OF THE
PRIMITIVE "LOVE IN A COTTAGE" SENTIMENT THAT LIVES IN
SOME DEGREE IN EVERY HUMAN HEART. ARCHITECTURALLY,
IT IS THE RESULT OF THE EFFORT TO BRING ABOUT
HARMONY BETWEEN THE HOUSE AND ITS SURROUNDINGS,
TO GET AS CLOSE AS POSSIBLE TO NATURE.
—*Radford's Artistic Bungalows*, 1908

The framer's name is Phil Borho. I first met him when he dropped by the site one day while the crew from Lake Huron Precast was putting up the foundation panels. I was down in the basement, helping them move 3½ tons of vertical concrete slab ¼ inch sideways so that the bolt holes would line up with those of its neighbour. We were using anything handy: a small crowbar, a wooden wedge, our feet. Most of the panel's weight was still supported by the crane, but it was a monumental task nonetheless. Just when we were feeling most foolish, I looked up, and there was Phil, standing on a pile of clay with the sun at his back.

Phil has been a framer for 30 years. Framers are carpenters who put up the shell of a house: the floors, the outside walls, the roof, the inside partition walls. They don't do finish work; almost nothing they do is visible when the house is complete, but what they do determines the final form as surely as a skeleton determines an animal's shape. They used to be called rough carpenters, but they have always referred to themselves as framers. Phil is a strong, fit man in his mid-50s, with a soft voice, callused hands and muscles around his right elbow well known to only carpenters and anatomists. Framers are methodical and prone to bemused speculation. Phil was looking down on the basement operations with a faintly sardonic expression, as if to say he'd seen a lot of changes, some of them good and some of them not so good, and he was beginning to suspect which category precast concrete panels would fall into.

When I climbed up to join him, I asked him how business was. "Pretty slow just now," he told me. Normally, he said, a carpenter knows he has houses to work on two or three months ahead. That might mean four or five houses lined up. "Now," he said, "you work house to house. You take what you can get." Things seemed to be picking up lately, he added: according to that day's *Kitchener-Waterloo Record*, 254 single-family houses had been sold in September, up 71 percent over the same month last year, "which shows you how bad things were last year."

92   Added to the bad news that fewer houses have been built in recent

years is the fact that they're getting smaller. Framers are paid for their work according to the size of the house – so much per square foot of floor space, not counting the basement. Because a 1,400-square-foot house doesn't take appreciably less time to frame than a 1,600-square-footer, a smaller house means less income for the framer. "I worked on a house two weeks ago that was only 900 square feet," he said, shaking his head. "It was smaller than some apartments I've seen. You just can't make a go of it anymore."

The trend toward smaller houses is a good one from an environmental standpoint, however. Since the turn of the century, average house size in North America has been going up almost as fast as family size has been going down. An average new house in 1912, for instance, was 1,500 square feet. During the Depression, the few houses that were built were small, under 1,000 square feet, but sizes began to climb again after the war, until in the 1960s, the average was 1,450 square feet; in 1989, it jumped to 2,000. I recently looked at two reputable magazines that advertise house plans which readers can order through the mail "at tremendous savings" and calculated the average floor area of the first 60 plans; it came to 2,250 square feet. The Greenhome officially has 2,500 square feet of living space, but that includes the basement. In the two magazines, basements were not included in the plans. If the Greenhome had been in one of them, it would have been listed as 1,250 square feet, the second smallest of the 60 houses in my sample.

In terms of the amount of space per occupant, the size of Canadian houses has gone way up. In 1912, the average family comprised 5.8 people; if they lived in an average-sized house for their time (1,500 square feet), each family member would have had 260 square feet of living space. Today, when the average family size is 3.1, even in a 2,000-square-foot house, each person has nearly 650 square feet. Before exploring the effects of unnecessarily large houses on the environment, let's ask ourselves why we feel we need so much space.

It's partly because we still think of the huge, rambling turn-of-the-century house as the ideal "home" – a large bay-windowed parlour, three storeys, ten bedrooms, high ceilings, back stairs, a summer kitchen, five chimneys and a wraparound porch. We like the style that such a house exhibits and associate the style with the size. What we fail to realize is that Victorians needed all that space; we don't. As Witold Rybczynski says, "The home of one hundred years ago meant something totally different to its inhabitants, and it contained a way of life unlike our own." Victorians had large, extended families, for instance, with half a dozen children, a maiden aunt or two, an uncle recently returned from the antipodes, aged grandparents, nannies and often servants, all living under one incredibly complicated roof. For them, home was the centre of their social lives; there were no cinemas, sports arenas, shopping malls, automobiles. Eating in restaurants was not a daily option. Only the adult male actually left home to go to work. In Victorian times, it was almost unheard of for no one to be at home. An empty house was a haunted house. Today, with both parents at work and the kids at school or in day care, we don't even *need* a house for most of our waking hours.

And yet we continue to make more room for ourselves. Big houses, like big cars, must fulfill some image we have of prosperity. Rybczynski is worth quoting at some length here, because he accurately pinpoints the link between locus and status: "Why," he asks, "do families that are on average smaller require twice as much space? To some extent, the expanding American house reflects a crude bigger-is-better mentality. Home ownership is a sign of social accomplishment and status, and just as the most prestigious cars were once the Cadillac and the Continental, which served as models for cheaper (but equally bloated) Fords and Chevrolets, the houses of the wealthy—in particular, Hollywood celebrities, whose sprawling Beverly Hills villas were prominently featured in fan magazines—were what the average tract house strove to imitate."

Perhaps big houses represent a kind of longing for earlier times, when families were bigger and closer. My own parents lived in small bungalows for as long as I can remember—my brother and I grew up in a suburb of Windsor, Ontario, in a bungalow that couldn't have been bigger than 700 square feet, including the front porch, and then in a slightly larger one in a suburb of North Bay—until both of us moved out and started having families of our own. Then my parents added a family room off the kitchen. Maybe we're all like that. Maybe we want large houses so that we'll be ready when the family comes home.

During the design meetings, Richard described the Greenhome as a raised bungalow "with the Santa Fe look," which means that the basement is partially lifted out of the ground and the main floor is ranch-style—sprawling and open but not too big. There was some talk about "the empty-nest syndrome": as the baby-boomer generation works its way through the economy like an agouti through a boa constrictor, more people are turning to smaller houses because the children have left home and parents are selling off the large family house and looking for something more manageable. Another rationale could include the fact that as more families split up, family size is decreasing while the actual number of families is increasing. The 1986 Canadian census showed that the number of families in Canada grew from 6.3 million in 1981 to 6.7 mil-

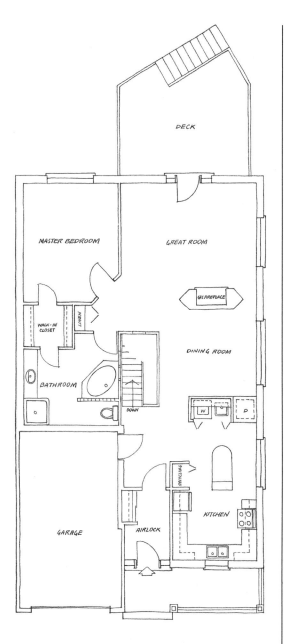

DECK

MASTER BEDROOM

GREAT ROOM

GAS FIREPLACE

WALK-IN CLOSET

LINEN

DINING ROOM

BATHROOM

DOWN

W

D

RECYCLING

GARAGE

AIRLOCK

KITCHEN

The Greenhome's main floor is designed on an open plan, with a freestanding gas fireplace dividing the great room from the dining room; the fireplace is direct-vented, and it alone is sufficient to heat the house. Modified cathedral ceilings provide spaciousness and improve air flow.

lion and that 2.2 million of those had no children at home. In other words, we don't need fewer larger houses, which is what we're building; we need more smaller houses like the Greenhome. Even Phil would be happy if he worked on more houses, as long as the total square footage of houses he framed each year remained the same.

The Greenhome is a bungalow in every sense of the word – it is small, detached, exurban and perfectly in tune with its environment – and whatever the actual size of the houses we've been building, or imagining, the history of housing in 20th-century North America has been the history of the bungalow. Although the form had humble beginnings, the bungalow has been the most successful house design in history.

Like "pajamas" and "verandah," the word "bungalow" comes from India: a *banggollo* was a 17th-century Bengali peasant's hut. During the time of the Raj, *banggollos* were taken over, often literally, by the English as temporary vacation houses; upon their return to England, these leisure-class folk built their own *banggollos* in the country or by the seaside, where they went to get away from smoggy Londontown. They were still temporary dwellings, simple and inexpensive houses built on land that wasn't otherwise being used.

The first bungalow to be given the name in North America was built at Monument Beach, Cape Cod, in 1880 by a Boston architect for a Boston banker. By the mid-1880s, everybody wanted one, and books and magazines of bungalow plans – defined as "medium- and low-cost houses" intended "as summer cottages" – were being mass-produced for America's rising middle class. Urban congestion produced a concomitant and largely artificial need to "get back to nature" – artificial because the nature people needed to get back to had to be within commuting distance of the city – and the bungalow was there to fill it.

The word "bungalow," however, referred more to the house's function as a country or vacation retreat than to its design. The first bungalows were quite a bit larger than the simply built structures we grace with the name today: many were two or more storeys high, with large rooms, high ceilings, tall windows and almost as much useless decorative detail as their city cousins. The back-to-nature people quickly realized that the seaside or lakefront was no place for multistoreyed, balustraded, crenellated, urban-style housing – that's what they were going there to get away from. The arts-and-crafts movement was born, devoted to simple construction and natural materials – rough-hewn joists, wood-panelled walls, cedar or redwood singles, fieldstone fireplaces. Bungalows were built by local craftspeople and in a style appropriate to the setting, standards that the Greenhome follows almost to the letter.

The bungalow was a very conscious return to the simplicity advocated by one of the first back-to-nature writers, Henry David Thoreau, who built his simple cabin—a 150-square-foot timber frame near Walden Pond—in 1845 and who advocated that a house ought to reflect the life of the occupant, not the whim of the architect: "The most interesting dwellings in this country," he wrote in *Walden*, "as the painter knows, are the most unpretending, humble log huts and cottages of the poor commonly; it is the life of the inhabitants whose shells they are, and not any peculiarity in their surfaces merely, which makes them picturesque; and equally interesting will be the citizen's suburban box, when his life shall be as simple and as agreeable to the imagination and there is as little straining after effect in the style of his dwelling." Thus a bungalow, wrote the editor of *The Craftsman* in 1906, was "a house reduced to its simplest form, where life can be carried on with the greatest amount of freedom; it never fails to harmonize with its surroundings." This meant low-pitched rooflines with large overhangs, horizontal construction, open verandahs and direct access to the outdoors from most rooms, features of which Thoreau would have heartily approved.

The back-to-nature ethos, with the bungalow as its central symbol, paved the way (as it were) for the next major development (so to speak) in house building, of which Thoreau would have disapproved: urban sprawl. Although cities in the east had been growing steadily, encroaching on their formerly sylvan environs from the 1880s on, the first glimpses of the true nature of rapid suburbanization came in the west. In Los Angeles, where land developers and electric-railway entrepreneurs parcelled up the land around the city and appealed to people in the east to come to California to escape the horrors of city life, the population soared from 100,000 in 1890 to more than one million by 1920. A similar explosion took place in Vancouver: in 1891, the city's population was 13,709; by 1921, after the completion of the railway and the opening of the Panama Canal, 222,294 people lived in the Greater Vancouver Regional District.

For the first time, the bungalow was not built as a vacation home but as a permanent residence, albeit in vacationland. The West Coast's benign climate made it possible to build inexpensive, flimsy houses with enough artistic flourishes to make them appealing to people who had gone there to escape the increasingly unaesthetic urban east. They weren't about to move into another downtown core. They wanted houses that reflected their new lease on life. The suburban boxes they bought up by the thousands were so perfectly suited to their environment that Mary Austin, a travel writer visiting California in 1914, wrote that bungalows were "as indigenous to the soil as if they had grown up out of it," as indeed many of them had, since they were built out of local rock, California redwood, B.C. Douglas fir and sand-textured stucco. The bungalow retained its sense of wholesome country living—simple, sensible and stylish—even though it had crossed a continent and nudged up against another inner city.

The bungalow with artistic flair didn't last long, however: by the 1920s, the word had passed into the language as a synonym for gimcrackery and shoddiness. When, in 1921, Woodrow Wilson accused President Warren Harding of being "bungalow-minded," he was saying not just that Harding thought small but that what he thought was cheap, mass-produced, insubstantial and tawdry. The image of the bungalow had sunk another notch in the social hierarchy: now it was lower-class.

But the dream of being a homeowner remained strong and was perhaps even fed by the architects' love affair throughout the 1950s with the high rise. To a prospective homeowner, however, moving from a high-rise apartment to a house would not have much appeal if the house in question were more shoddily constructed, more cramped and less expensive than the apartment. Builders of suburban bungalows during the boom in the 1950s, when the concept of suburbia really took hold in almost every city on the continent, had to make houses that were bigger, more attractive and still barely affordable. They invented curb appeal and cost-cutting construction techniques; they lobbied for greater density regulations and looser building-code restrictions. Moving into the suburbs in the 1950s was a sign of having made it. The bungalow became middle-class again. When we watched shows like *Father Knows Best* and *Leave It to Beaver* on television, we were imbibing more than the values of the all-American family; we were also absorbing an image of the all-American home. And it was in Scarsdale, or Scarborough. The conspicuous prosperity fostered its own symbols—dishwashers, large refrigerators, huge hi-fis—all of which required more space.

But Ward and June Cleaver never had to worry about oil shortages and ozone depletion; their chief concern was to see to it that the Beave made it through school, started his own family and bought an even bigger house than theirs. Times have changed. The back-to-nature movement that spawned the bungalow in the 1920s resurfaced as the back-to-the-land movement of the 1970s. Unfortunately for the bungalow, it had become a symbol of what the back-to-the-landers were trying to avoid—Scarberia—and so the early environment-conscious builders turned to huge, high-tech designer houses.

98 Times have changed again. Smaller houses make environmental sense,

which surprisingly often is nothing more than common sense. For one thing, smaller houses are easier to heat; for another, they take up less land. And developers as well as owner-builders know they are less expensive to build, because they use less material: in 1990, according to the Canadian Wood Council, an average single-family house in Canada used 10,000 board feet of lumber—or about one acre of softwood forest. And as the Greater Toronto Home Builders' Association (GTHBA) has pointed out in a booklet called *Making a Molehill out of a Mountain*, about 10 percent of that—the equivalent of 200 2-by-4 studs—is waste. The GTHBA was concerned about the dollar cost that waste represents: "Since 1983," the booklet says, "tipping fees alone have risen over 600 percent. . . . On average, builders are paying $300 per home to dispose of approximately 2.5 tonnes of garbage." But the environmental costs are far more serious than the economic ones: multiply that 2.5 tonnes by the 35,000 new houses built in the Toronto area each year, and you have the equivalent of 3,500 acres of softwood forest being dumped into landfill sites every year just in the Toronto area.

No construction waste from the Greenhome is going to a landfill. What waste there is will be shipped to a recycling depot or reused on another job site, just as, a century ago, leftover lumber from a framing bee was either taken home by a neighbour or used for kindling. Once again, the Greenhome is taking a step forward by taking a step back.

□ □

When Phil is ready to start framing, he finds problems with the foundation that have to be corrected before he can begin. "It bows out," he says, standing at the southwest corner and sighting down the length of the south wall. "Look at that." He walks along the outside of the wall, running his right hand across the precast panels as if stroking the flank of a bloated horse. "See, here in the middle, it's at least an inch out of plumb." Then he sights along the top of the wall. "And it's humped up in the middle too," he says, shaking his head. "Looks like that centre pad's about three-quarters of an inch high."

Other problems turn up. The north and west walls bow out as much as the south wall. The opening for the downstairs walk-out is too narrow. The notch in the concrete for the girder that will support the main floor is in the wrong place. All of these things are correctable, but they are annoying and will cause delays. "If one of my own crews handed me a foundation like this," says Werner, "I'd fire them."

What turns out not to be correctable are two concrete panels that form the inside walls for the garage. Think of the foundation, seen from above, as one large rectangle divided into nine smaller rectangles; the top right rectangle is the garage. The two interior walls of the garage foundation have been installed backwards, so the waffled sides face into the garage rather than into the house. To turn them around now would require the crane, which would cost a minimum of $500 to get back to do a half-hour job. Werner is inclined to call the crane in anyway and deduct the amount from the $8,000 Lake Huron Precast is charging for the foundation, but he decides to check with John Kokko at Enermodal first to see whether there's a less expensive alternative. He walks off to his truck muttering to himself, while Peter and Bert, looking sheepish, set about fastening a pair of come-alongs to the inside of the walls across the middle so that they can pull them in from the top and correct the bows. Phil takes his chalk line out to snap a line along the top of the south wall to find out exactly how much it humps.

"A bowed foundation wall is not all that unusual," says Ken, Phil's son and one of his two assistants. "It sometimes happens that a foundation is backfilled too early, before the concrete is completely set, and the wall is pushed inward by the weight of the earth against it. Then we just have to dig it out again and straighten it, as long as we catch it before the concrete hardens." The hump in the south wall isn't that unusual either. Jim Locke, in The Well-Built House, warns that even with poured-in-place concrete, the tops of the plywood forms can get hacked up from being used so often, imparting their imperfections to the top of the wall. Locke generally orders his foundation walls an inch or so short of 8 feet so that the concrete is not poured up to the top of the forms. Uneven foundation tops can be corrected with shims, but if the wall is almost an inch out, as this one is in places, shims are impractical, creating air leaks between the plate and the concrete that would require hours of caulking to correct. Phil decides that the best thing to do is to rip the 2-by-6 sill plate so that it sits snugly over the hump. The top of the plate will be level even though it will be thin where the hump is highest.

In traditional poured-concrete construction, J-bolts are embedded in the top of the wall every 2 or 3 feet, and the sill plate is bolted down, sandwiching a weatherproof gasket between the wood and the concrete. In the case of precast panels, holes are left in the top of the panel at 24-inch intervals, 12-inch bolts slide down into them through holes drilled into the sill plate, and nuts and washers are applied from below. No gasket is required, because the whole house will be clad in a continuous external sheathing extending from soil to soffits, covering the seam.

When the bowed-out walls have been corrected by the come-alongs

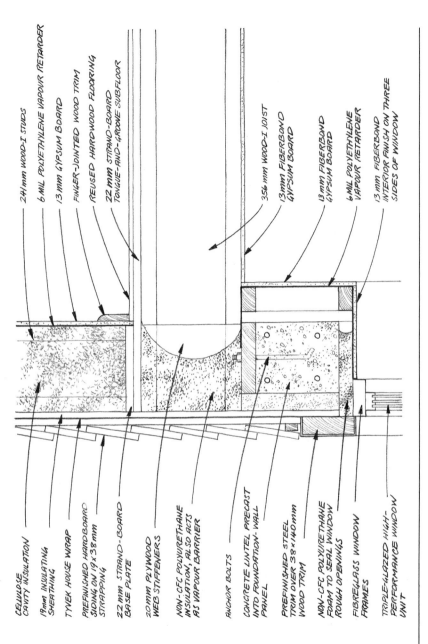

Labels (reading clockwise from top):

- CELLULOSE CAVITY INSULATION
- 19mm INSULATING SHEATHING
- TYVEK HOUSE WRAP
- 24mm WOOD-I STUDS
- 6 MIL POLYETHYLENE VAPOUR RETARDER
- 13mm GYPSUM BOARD
- FINGER-JOINTED WOOD TRIM
- REUSED HARDWOOD FLOORING
- 22mm STRAND-BOARD TONGUE-AND-GROOVE SUBFLOOR
- 356mm WOOD-I JOIST
- 13mm FIBERBOND GYPSUM BOARD
- 13mm FIBERBOND GYPSUM BOARD
- 6 MIL POLYETHYLENE VAPOUR RETARDER
- 13mm FIBERBOND INTERIOR FINISH ON THREE SIDES OF WINDOW
- TRIPLE-GLAZED HIGH-PERFORMANCE WINDOW UNIT
- FIBREGLASS WINDOW FRAMES
- NON-CFC POLYURETHANE FOAM TO SEAL WINDOW ROUGH OPENINGS
- PREFINISHED STEEL TRIM OVER 38×140mm WOOD TRIM
- CONCRETE LINTEL PRECAST INTO FOUNDATION-WALL PANEL
- ANCHOR BOLTS
- NON-CFC POLYURETHANE INSULATION, ALSO ACTS AS VAPOUR BARRIER
- 20mm PLYWOOD WEB STIFFENERS
- 22mm STRAND-BOARD BASE PLATE
- PREFINISHED HARDBOARD SIDING ON 19×38mm STRAPPING

Blown-in cellulose insulates the foundation and first-floor wall cavities; the floor ends are filled with a non-CFC-blown polyurethane foam that also acts as a vapour retardant. The exterior siding and most of the interior wood is made of recycled wood fibre and oriented-strand board.

101

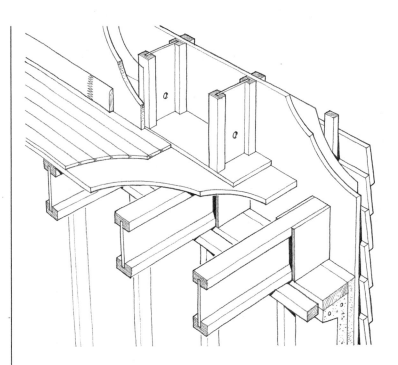

The Greenhome's floor joists, wood-I beams made by Trus Joist MacMillan, consist of 2-by-4s joined by aspenite webs. They are stronger than traditional 2-by-10s and use one-third less lumber. The subfloor is ⅞-inch aspenite, and the floor is hardwood salvaged from a nearby distillery.

and all the sill plates are on, Phil, Ken and Ted begin to lay the floor joists. The system the design team has chosen is known as the Trus Joist Silent Floor, and although it has been around for a number of years, it still qualifies as an innovative floor system. (Trus Joist Corp. began making it in 1969, but it has only been generally available since Trus Joist merged with MacMillan Bloedel Limited in 1991.) Normally, floor joists are a series of 2-by-8s, 2-by-10s or even 2-by-12s that span the foundation walls, one end of each joist resting on the sill plate and the other on a girder that runs down the middle of the house. A Trus Joist works the same way, except that instead of being a 2-by-10 on its edge, it is a kind of wooden I-beam (called a "wood-I" joist) made by joining two 2-by-4s with a strip of plywood. The 2-by-4s are dadoed, or grooved, down the centre of their widths, and a plywood "web" is pressure-glued into the dadoes to complete the I.

The system has several advantages over solid-sawn lumber: it uses only small dimensional pieces (2-by-4s), thus saving trees; the exact length of joists required can be ordered from the factory, so there are almost no off-cuts left at the building site; and the joists have more strength than dimensional lumber, which means they can be placed over a wider span and farther apart. These Trus Joists are placed 24 inches apart and run right across the width of the house, with only a single, off-centre supporting wall under them. If the joists had been 2-by-10s, each would have been made by lap-nailing three 2-by-10s together over two girders, and they would have had to be placed on 16-inch centres. In other words, to span a 4-foot section of the Greenhome floor, the Trus Joist system used six 2-by-4s instead of four 2-by-10s.

To express the savings another way, a 1,250-square-foot floor such as the Greenhome's ordinarily requires 280 cubic feet of 2-by-10 solid-sawn lumber; using 14-inch Trus Joists on 24-inch centres, the Greenhome used less than 90 cubic feet of lumber. According to Trus Joist MacMillan, it takes 3¾ hypothetical trees 120 feet tall and 20 inches in diameter to supply 280 cubic feet of 2-by-10s, whereas only 1¼ such trees are needed to make 90 cubic feet of Trus Joist lumber. Even better, the 2-by-4s used in Trus Joists don't need to come from 20-inch-diameter trees; they can come from smaller, younger trees planted for the purpose.

This is good news, because there aren't that many big trees left. It has recently been estimated that of the 70,000 square miles of forest that existed in North America at the time of Columbus, only about 10 percent is still standing as old-growth timber; 7,000 square miles, or about 4.5 million acres, is all that's left of trees of the size that can produce 2-by-10 or 2-by-12 lumber. Although the forest industry is busily planting fast-growing varieties of replacement trees, the American Forest Council still warns that about 1 percent of the remaining old-growth forest is falling to commercial harvesters each year, which depletes the availability of—and therefore increases the price of, and therefore enhances the desirability of—old-growth lumber. The ability of engineered lumber such as wood-I joists to fulfill the roles traditionally assigned to large dimensional lumber will go a long way toward protecting those bigger trees. It seems that this is already starting to happen. A 1990 survey of 3,500 American building suppliers found that the 170 million linear feet of wood-I joists manufactured by five companies in 1988 is expected to rise to 318 million by the year 2000—about 19 percent of the market. This means that one house in five will be built with wood-I joists. No wonder MacBlo was keen to join forces with Trus Joist Corp.

Another floor system similar to Trus Joist had been brought up at the design meetings: the SpaceJoist System, manufactured by Truswal. These engineered joists use regular "on-flat" 2-by-4s, but instead of being connected by a strip of plywood down the centre, as Trus Joists are, the 2-by-4 chords are joined on each side by galvanized steel V webs. Teeth at each point of the V are machine-pressed into the side of the wood. SpaceJoists are engineered for 40-foot spans, and the open webbing leaves lots of space for ductwork and electrical wiring, but John was reluctant to use any more steel in the Greenhome than necessary: "After CFCs," he said, "the metal industry is the second biggest enemy of the ozone layer. Holes for ductwork can be cut in the plywood webs in the Trus Joist system, so I don't really see the advantage of using SpaceJoists."

Phil, Ken and Ted lift the Trus Joists into place. All the joists were custom-made at the factory. When they arrive on site, the end of each joist is coded with a letter: A and D for the longest spans at the back of the house; B and C for the shorter spans, where the joists meet the stairwell leading down to the basement; E for those stretching from the south wall to the garage wall; and F for the shortest, 2 feet long, that tie the end joist to the foundation wall. When they are all in place and their levels checked, Ken starts toe-nailing them down while Ted begins applying the cross bridging, Xs of metal strips nailed from the top of one joist to the bottom of the next over the sill plate to keep the joists vertical. (Regular 2-by-10 joists require bridging every 4 or 5 feet along the length of the span; because Trus Joists literally cannot warp, bridging is needed only at the ends.) Phil cuts a number of 14-inch plywood squares that will be nailed to the end of each joist as "cripples" to give extra support under the load-bearing joist ends. Ken nails these on with the Paslode hammer gun attached to the end of a red airhose that runs to a compressor by the truck. The compressor is noisy, and when Werner returns, he and Phil have to shout above it. The previous night, both Werner and Phil stayed up late to watch the Blue Jays lose the third game of the American League Championship Series to the Oakland A's. Both Werner and Phil feel strongly that the Jays lost because their manager, Cito Gaston, left the starting pitcher, Jack Morris, in too long.

"What's he saving the bull pen for?" asks Phil loudly. "Next year?"

"He's been doing the same thing all season long," says Werner. "I think Morris has a clause in his contract saying he can't be pulled as long as he has a chance of winning."

Phil has worked a lot for Werner. They like and respect each other. "Phil is a good framer," Werner told me back when the hole was being excavated. "He's quick and methodical, he doesn't complain, he just gets on with it. I can leave him on a job and not have to worry if it's being done right."

"Werner's a good contractor," Phil told me on a different occasion. "He'll wait for me. If he calls me up for a job and I tell him I can't get there for a week or so, he'll wait. A lot of other contractors would just call someone else."

"Did your wife stay up to watch the ninth?" Phil asks Werner.

"No," says Werner. "She fell asleep just before Baines got his home run."

"She's lucky, then."

Werner nods. "I spoke to the architect about those garage panels," he says, getting at last to the point of his visit. "He says we can leave them the way they are and build an inside stud wall to hold the insulation. We'll fill the cavities on the outside with fibreglass and put some rigid board against it before we backfill. It's not a great solution, but it'll do."

Phil laughs and shakes his head. "Either those panels are backwards," he says, "or the architect is."

The decking that comes with the Trus Joists is not plywood but ⅞-inch aspenite sheeting, a form of chipboard (technically called oriented-strand board, or OSB) with one side slightly roughened so that the deck won't be slippery to work on when wet. Usually, the subflooring is ½-inch spruce plywood. Aspenite has two advantages: first, its extra thickness means that it won't sag between joists which are 2 feet apart, and second, aspen is a more environmentally responsible product than spruce. In the natural boreal forest, hardwoods like aspen and poplar grow with such conifers as spruce and pine; the young conifers need the faster-growing aspen for protection against frost and weevil damage. Aspens, though, were shunned by the pulp-and-paper industry because their fibres are too dense to make newsprint, so they were kept down with herbicides, and at clear-cut time, they were left to rot on the forest floor.

The long-term effect of such measures, however, is that what used to be a nicely mixed boreal forest has been growing back with a preponderance of aspen and birch—what used to be termed "junk trees" because they had no commercial value. According to Jerry Franklin, chief plant ecologist with the U.S. Department of Agriculture's Forest Service, "most forests in the temperate zone are secondary forests that developed after logging of primeval forests or abandonment of agricultural lands. The composition and structure of these forests are different—often drastically different—from those they have replaced. We see, for example, forests of birch and aspen in the Great Lakes states, where the forests were originally dominated by long-lived pioneer species such as red and eastern white pine and late successional species of hardwood." As a result, the industry has taken a second look at junk species and realized that aspen is not such a bad tree after all. In fact, more and more aspen is finding its way into pulpwood destined for the paper industry, and new construction uses are found for it every day. At one time, because the different maturing rates for aspen and spruce complicated life for the harvesters, forest-management people considered growing them separately on monocultural "fibre farms" – aspen in this woodlot, spruce in that one – but pressure from environmental groups concerned about biodiversity has convinced the industry that selective harvesting, once deemed not cost-effective, is now in fact the answer to several problems. Aspen can be harvested for chipboard after about 20 years, by which time the spruce have outgrown their need for frost and weevil protection; in the meantime, forest wildlife has a healthy mixed forest to dwell in, and there is no need for extensive (and expensive) herbicide campaigns.

Aspen is also the principal ingredient in the Greenhome's walls, which are made from wood-I trusses similar to but smaller than the floor joists. In this case, the 9½-inch wall trusses are made from parallel-strand lumber (PSL), marketed under the names Parallam and PSL 300, again by Trus Joist MacMillan. PSL is the engineered lumber product that MacMillan Bloedel brought to the Trus Joist union. Developed in the early 1980s, PSL has won MacBlo the 1984 Canada Award for Excellence, the 1987 Science Council of British Columbia Gold Medal and Sweden's Marcus Wallenberg Prize. It consists of the same kind of veneer that is "peeled" from logs for plywood, but instead of being used whole, the veneer is cut into long strands and glued into lengths with a formaldehyde-free resin: the lumber is literally extruded from a rotary belt press in a continuous billet, like square sausage from a sausage machine, and cut off at any length up to 66 feet. This billet is then sawn like a log into dimensional lumber – 2-by-2s, 2-by-4s, 4-by-8s, whatever is wanted. The girder under the Greenhome floor joists is a 12-foot length of 6-by-14 PSL, and the wall trusses are made by joining two PSL 2-by-2s with a strip of aspenite.

Low-energy builders have been trying for decades to get thicker walls that provide more room for insulation. Even with modern insulation, which has greatly enhanced R-values per inch, traditional 2-by-4 stud walls just don't have enough room to superinsulate a house. For a while, builders were making two 2-by-4 walls, one inside the other, called a double wall, for an overall thickness of 7 inches; then they made a double wall consisting of one 2-by-6 wall and one 2-by-4 wall, with the vapour barrier between them. This system, though it provides adequate room for insulation, simply requires too much wood – the equivalent of a 2-by-10 wall around the outside of the house. Traditional builders couldn't afford it, and environment-conscious builders didn't want to

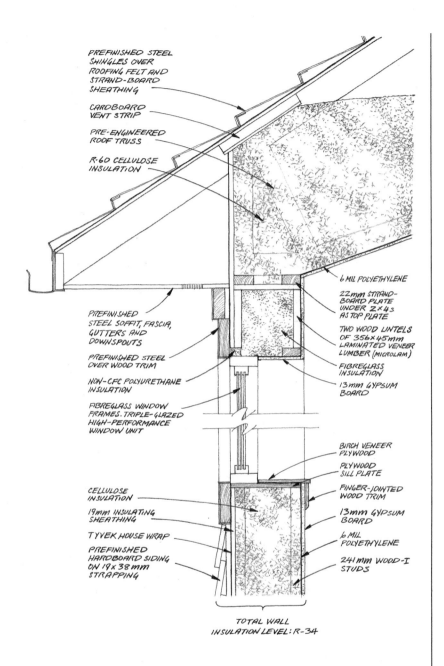

PREFINISHED STEEL SHINGLES OVER ROOFING FELT AND STRAND-BOARD SHEATHING

CARDBOARD VENT STRIP

PRE-ENGINEERED ROOF TRUSS

R-60 CELLULOSE INSULATION

PREFINISHED STEEL SOFFIT, FASCIA, GUTTERS AND DOWNSPOUTS

PREFINISHED STEEL OVER WOOD TRIM

NON-CFC POLYURETHANE INSULATION

FIBREGLASS WINDOW FRAMES. TRIPLE-GLAZED HIGH-PERFORMANCE WINDOW UNIT

CELLULOSE INSULATION

19mm INSULATING SHEATHING

TYVEK HOUSE WRAP

PREFINISHED HARDBOARD SIDING ON 19x38mm STRAPPING

6 MIL POLYETHYLENE

22mm STRAND-BOARD PLATE UNDER 2x4s AS TOP PLATE

TWO WOOD LINTELS OF 356x45mm LAMINATED VENEER LUMBER (MICROLAM)

FIBREGLASS INSULATION

13mm GYPSUM BOARD

BIRCH VENEER PLYWOOD

PLYWOOD SILL PLATE

FINGER-JOINTED WOOD TRIM

13mm GYPSUM BOARD

6 MIL POLYETHYLENE

241mm WOOD-I STUDS

TOTAL WALL INSULATION LEVEL: R-34

The 9½-inch exterior walls are wood-I trusses made of parallel-strand lumber and aspenite. The exterior sheathing is a new product called Excel Board, made of non-CFC-blown polystyrene mixed with recycled wood fibres.

cut down so much forest. A wall system using PSL trusses requires half the volume of wood that a 2-by-6 stud wall does and about one-third the amount needed for a double wall. There is no waste, warping or shrinking, and it provides a larger wall cavity for insulation.

At the meetings, Ian was worried that the wood-I wall trusses would have problems with thermal bridging, which means that the insulation values between the trusses would be all right – the design called for R-34 – but that the trusses themselves would conduct cold from the outside to the inside, lowering the wall's overall efficiency. Back at the Enermodal office, John's colleague Alex McGowan ran a computer check on the system. He fed in a model with 7¼-inch trusses, ½-inch drywall on the inside, ordinary clapboard siding on the outside and an insulation value of R-34 between them. He invented an outside temperature of minus 18 degrees C and an inside temperature of 18 degrees and asked the computer what the temperature would be on the inside surface of the drywall where it met the truss; the computer said 17.3 degrees. No appreciable thermal bridging. And the difference would be even less with the 9½-inch trusses in the Greenhome.

Phil lays the exterior walls flat out on the decking, one truss every 24 inches, and power-nails two strips of plywood along the bottom – one ½-inch strip and one ⅜-inch strip – to give a ⅞-inch bottom plate. (The plans call for these strips to be cut from leftover aspenite decking, but there is no leftover decking. And the trusses were precut before delivery, so Phil can't just make them ⅜ inch longer, hence the double bottom plate: plywood doesn't come ⅞ inch thick.) He uses ½-inch plywood for the top plate. When the walls are assembled and nailed together on the deck, he and Ken stand them up and lift them into place at the edge of the deck, bracing them vertical with 2-by-4s, exactly as they would have with ordinary stickwall construction. Later, when the cladding (which I describe later in this chapter) is nailed to the outside of the wall, the interior of the Greenhome looks like a huge open-air library with the vertical shelving members up but the shelves removed.

"I liked working with the floor trusses," Phil says. "They're light and easy for one man to set in place, they don't require a lot of strapping, and with the 2-by-4s laid flat side up like that, there's twice the nailing surface for the decking. You don't have to worry so much about missing with the nailer. And they're incredibly solid. To be honest, when I saw the span they were supposed to cover, I was expecting the floor to bounce up and down like a trampoline. But it's the most solid floor I've ever built. No give to it at all," he says, jumping up and down without lifting his feet off the floor to demonstrate his point. There is no bounce. "Even the best 2-by-10 floors have some give. And I also like this aspenite deck sheeting," he says. "With the rain we've been getting, I expected to find the deck swollen up with water today. But when I got here, there was about an inch of water sitting on it, and in most places, it was still ⅞ inch thick. No swelling, except at the walk-out. If we'd painted that edge last night to keep the water out, there wouldn't be any swelling at all. Ordinary plywood would've warped and twisted and come unlaminated, but not this stuff."

When the exterior walls are up, Phil paces out the floor with a tape measure and a chalk line, the architect's plans unfurled in his hands, figuring out where the interior walls will go. In an hour, the aspenite floor is zigzagged with blue chalk lines—kitchen here, bedroom over there, hall closet here—and Ken and Ted are busy with the power gun, making stud walls on the floor to be lifted into place along the chalk lines. The interior walls are the only place where traditional 2-by-4 spruce or fir studs are being used. There is plenty of second- and third-growth spruce and fir of a size suitable for 2-by-4s, and solid-sawn 2-by-4s have about the same embodied energy as engineered lumber of the same dimensions and about one-ninth that of steel studs, which so far are not being made from recycled steel.

By the end of the day, October 16, most of the interior walls on both floors are up, and a cold, light drizzle is setting in. Standing in the middle of what will be "the great room," modern architecture's word for the living-and-dining area, Phil looks up at the greying sky. The rain is only a degree or two from being snow.

"A roof would be nice," he says.

□ □ □

The simplest of all possible roofs is an inverted floor. Flat roofs are practical in hot, dry climates, because they do not deal overmuch with water and they do not create lofts that entrap heat. In climates where rain and snow are present in significant quantities, however, a flat roof loses almost all its appeal, a fact that does not prevent engineers from continuing to try to make them work. When I lived in Montreal, where the annual precipitation is measured in feet, I used to marvel at how many two-storey houses had flat roofs—and at how many upstairs ceilings had tea-coloured water stains spreading through the plaster.

The next simplest roof is the shed roof, which is just a flat roof tilted to shed rainwater, hence its name. Like Lord Davenport, it has lent its name to the structure it most often sits on. Originally, a shed was nothing more than a single-sloped roof on four corner posts used as a shelter for animals. Shed roofs have been used on more elaborate structures than sheds, of course, but as buildings increased in size, the spans required for the rafters began to assume unmanageable proportions, and some other system had to be devised. Even where long logs are available, shed roofs sag under heavy snow loads; a friend who has a cathedral-ceilinged shed-roofed house in the Laurentians, north of Montreal, has had more major work done on her roof than on any other part of her house, and it still sways in winter like the back of an overladen packhorse.

The answer to one long shed roof is two shorter shed roofs placed back to back and joined at the high point by a ridgepole. This is called a gable roof (the gable is the triangular bit of the house wall; the word is related to Old Teutonic *gathla*, meaning "fork," which reminds us that ridgepoles were originally called rooftrees) and is the most common roof found in North America. Even gable roofs sag under heavy snow loads, however, and the downward pressure on both angles of the roof tends to push the walls out. This problem was attacked by adding roof joists that ran from wall to wall to tie them together and complete the bottom of the triangle. Throwing a few boards over the joists also provided extra storage space for all those valuable things I now have in my basement; at the same time, it conveniently made a ceiling.

For really heavy snow loads, however, simply adding joists wasn't good enough. Joists didn't stop the sag, they just prevented the sag from pushing the walls apart. Making the rafters out of larger and larger dimensional lumber was one answer, as long as the house was small. Thoreau describes a loggers' shanty in Maine made out of fresh-cut pine logs, with the gable ends tapered by using shorter and shorter logs right up to the peak and then laying more logs lengthwise from gable to gable so that the roof was essentially a continuation of the wall. For more permanent structures, however, a less rough-hewn solution had to be found. The problem was to transfer the weight of the snow so that it bore downward, not outward. The solution was to invent the roof truss.

Extra roof support was provided by adding a "king" post running vertically from the centre of each joist straight up to the ridge of the roof. In larger structures, two shorter posts (called queen, or side, posts) ran up from each joist to about the midpoint of each rafter, dividing the joist into thirds. Struts joined the bottom of the king post with the top of the queen post. Each unit—two rafters, one joist and one, two or more posts and struts—could be built separately on the ground and hoisted up to the roof; it was called a truss.

As trusses became more and more complicated, and as the supply of large, old-growth lumber dwindled, houses became more and more expensive. For a long time, we kept on building trusses anyway, because it was, well, the way we'd always done it: roof joists of 2-by-10s, rafters and posts of 2-by-8s, struts of 2-by-4s. Eventually, truss lumber became so heavy that just nailing a post to a rafter wouldn't work. Simple bolts were tried, and in 1933, the United States Department of Commerce's National Committee on Wood Utilization started researching more elaborate metal fasteners.

The lumber industry was soon introduced to something called TECO timber connectors (TECO stood for the Timber Engineering Company). The best known of these were called split rings; the ends of the trusses were hollowed out and metal plates inserted, bolted to the trusses and connected to large metal plates that were bolted to the rafters at one end and to the joists at the other. This incredibly unwieldy system—roughly the equivalent of building a railway bridge in the attic—was used extensively until after World War II, when various companies began experimenting with nailed plywood truss plates and, eventually, galvanized steel plates. The advantages of the latter quickly established them as the norm, and a dozen companies sprang up to fill the growing demand.

These companies also supplied engineers' drawings for different applications of the truss-plate system, depending on size of wood, amount of snowfall, roof span, and so on, and when the fad for prefabricated houses hit in the 1960s, the idea of prefabricating roof trusses caught on rapidly. After all, if 9 out of 10 roofs were going to be gable roofs, why not standardize the pattern and prefabricate the trusses? By the mid-1970s, most building-supply companies were selling prefabricated roof trusses—each truss consisting of a joist, two rafters and as many trusses as local snow conditions required, all engineer-designed and approved—and all a builder had to do was figure out how many were needed and order them.

The benefits to the building trade were enormous savings in time and labour as well as an increased efficiency in the ability of a roof to transfer weight. But the benefits to the environment are equally significant. As with their modern counterparts—wall and floor trusses—prefabricated roof trusses allow builders to use less wood: on average, 1,000 board feet less per house. And lumber of a smaller dimension: in the trusses built for the Greenhome, for example, there is no piece of wood bigger than a 2-by-4, and many of the struts and posts are 2-by-3s.

Still, there is a lot of wood in the house. The great room and the master bedroom both have arched cathedral ceilings, and each of the trusses above them is made with no fewer than 24 separate pieces of lumber. The trusses over the kitchen, which has a flat ceiling, each contain 19 pieces. While Phil, Ken and Ted hoist them into position across the top of the walls and crawl in and around them, nailing them into place, I stand on the floor and look up. I seem to be peering through a forest of lumber. It's as though I'm standing on the deck of a sailing ship, looking up, like Robert Herrick, at sailors climbing in spars. But this impression is short-lived: by the end of the day, the trusses are in place, on 24-inch centres, 1-by-6 spacers toe-nailed between them at their tips—diminutive remnants of the mighty rooftree—and 96 sheets of waferboard roofing have been air-nailed over them, the noise from the hammer echoing through the now-enclosed space. With the view of the grey sky cut off, the house no longer feels like a ship; it feels like a barn.

□ □ □ □

John has been on the phone about the doors. The Greenhome will have six insulated doors when it's finished: a front door; a door leading from the airlock entry to the heated part of the house; another going into the unheated garage from the front hall; a rear door leading out to the deck from the great room; another on the lower floor walk-out; and one in the downstairs bathroom leading to the fruit cellar. The problem with exterior doors, from the Greenhome perspective, is that almost all of them are metal on the outside and insulated with foam cores whose manufacturing process uses CFCs. John has been trying to find fibreglass doors that do not use CFCs. When I get back to his office, I find him sitting with his elbows on his desk and his head in his hands.

John was an automobile mechanic before he turned to mechanical engineering; in 1981, he volunteered for Pollution Probe, helping them build Ecology House, "mostly providing brawn, not brain," he says. It was while working on that house that he realized he was in the wrong business. "I kept looking at the blueprints and thinking, If I had some education, I could probably make as many mistakes as these engineers did." So he went to Ryerson Polytechnical Institute in Toronto, got his engineering degree and eventually returned to Ecology House as technical coordinator, looking after the mechanical systems.

To earn a living, however, he went to work for Allen, Drerup and White, who by then owned a company in Toronto called Air Changer, which manufactured heat-recovery ventilators. When ADW sold the company in 1985, he came to Enermodal to help Steve perfect his heat exchanger. His commitment to the environment is so complete that he has built it into his definition of comfort: a house, he says, has to be comfortable, but he could not be comfortable in a house that was not energy-efficient. As a way of showing how far such a commitment has taken him, let me pass on the story of John's Volvo.

"I was driving a Volvo," he says, "which I thought was reasonably okay, but when I started really looking at it and looking at the gas mileage other cars were getting, I started feeling guilty about driving the thing. So I decided to get rid of it. But I couldn't just trade it in, because then someone else would be driving it around, burning up gas and polluting. I couldn't sell it for parts, because then it could be used to keep 10 Volvos on the road. In the end, I decided to purposely ruin the engine; I filled the cylinders with sand and put a brick on the gas pedal and just left it until it stopped running. Then I sold it to a scrap-metal dealer so it would be recycled." This from a man who says, "I've never thought of myself as a radical."

With this in mind, John's philosophical agonizing over various aspects of the Greenhome's efficiency, so irksome to the more practical-minded Werner, begins to make sense. He never sees a material as a thing in itself but rather as the midpoint in a process that began in the Earth and will end in the Earth. Take nails, for example. He worried the nail question for days before the house was even started. Nails are made of steel, and the steel industry is the second-biggest contributor to global warming and ozone depletion after the Freon industry. But even John paled at the thought of telling Werner and Phil that they'd have to build the house without nails. What could they do, go back to using wooden pegs? John probably considered it. But there just isn't a viable alternative to steel nails. There is, however, a viable alternative to steel doors.

"Fibreglass," he says. "We were going to use Stanley doors, which are made from metal and have a water-blown polyurethane-foam core. But I've heard they have had to replace half a million dollars' worth of doors recently because shrinkage in the foam core was warping the doors. I've also heard that when polyurethane shrinks, it loses quite a bit of its insulative properties, even if it doesn't warp the door. So I called the guy in Toronto who was going to install them for us, and I asked him about it. He says he's got something 'almost as good' as water-blown foam-core doors—a door with 50 percent reduced CFCs. I told him 50 percent reduced is not 'almost as good' as 100 percent reduced." It was like telling John that a Volvo is almost as good as a Toyota because it gets half the gas mileage. "I told him I'd keep the door open on doors and look around for another manufacturer. He said he knew of 58 different door manufacturers in North America, and none of them used foam cores without CFCs. So when I got off the phone, I looked through our catalogues and found one in about five minutes that makes a fibreglass door with a non-CFC-blown polystyrene core. It's a company called Pease, from Ohio. I liked the sound of that—Pease. So I called them, and they said sure, they'll ship us six doors. They haven't decided whether they'll donate them or sell them to us, but even if they don't donate them, I think we'll buy them. At this point, we can't afford to mess around."

The choice of windows was a bit easier, because Enermodal has been testing windows for a decade and knows what's what and where to get it. Since it's Enermodal's computer program that gets a product on Ontario Hydro's list of energy-efficient windows, all the design team had to do was go to the list and choose the window with the highest energy rating. It turned out to be manufactured by Accurate Dorwin Co., a Winnipeg firm, using components from a number of other companies in Canada and the United States.

"The energy rating," says John, "is simply a calculation based on total solar-heat gain through the window, minus heat loss by convection, conduction and infiltration out through the window. It's given in watts per square metre, prorated over the entire heating season." For a while, Ontario Hydro was awarding $5 per square foot of window to any new homeowner installing windows with an energy rating greater than minus 11 (only if the new homeowner also happened to be heating with electricity, of course), which means the window loses less than 11 watts per square metre of glass. The Dorwin window has an energy rating of plus 1; a Dorwin window will actually *gain* 1 watt per square metre of window area each winter.

"The numbers get a bit fuzzy," says John, "because we have to average them out to include each of the four cardinal compass directions and factor in the average weather conditions across Canada—solar radiation, heating degree days, and so on—so that the rating applies anywhere in the country, no matter where it's installed or which way it's facing. But essentially, it means that in any given place in Canada, this window will outperform any other window on the market."

Dorwin's windows are superwindows. Since the early 1980s, most new windows installed in new houses have been double-glazed sealed units in aluminum or wooden frames. These are not superwindows; they are only minimal improvements over the old single-glazed sash windows with sliding aluminum storms that, in my experience, can be jacked open only a few inches in summer and won't slide back shut at all in winter. Double-glazed windows with aluminum frames have an R-value of about 2, which, by the way, is the insulative value measured at the exact centre of the window, as far away from the frame as possible, and is therefore highly misleading in terms of the window's overall efficiency.

The problem with ordinary glass, whether in single- or double-glazed windows, is that it lets more heat out of the house than it allows in. Sunlight, or solar energy, enters through the glass and heats up objects inside the house; radiant heat emitted from those objects hits the glass from the inside and moves through it by conduction or, if the window is leaky, around it by convection. You can seal the window by caulking around the frame to cut down convection losses, and you can cover the glass at night with pretty quilted fabric to reduce conductive losses, but all in all, you're still going to lose more heat through regular double-glazed windows than you gain over the entire heating season, especially if the windows face north, because they don't gain much solar energy during the day. No matter how you cut it, the windows in a typical new house account for up to 35 percent of its total heat loss. Until the mid-1980s, this was the rock on which the solar-energy ship foundered.

In 1982, a California manufacturer, AFG, introduced a concept that revolutionized the window business and put solar energy back into safer waters. The concept was called low-E coating; low-E stands for low emissivity. The glass in a low-E window is coated with an invisible micro-thin layer of metallic oxides that bounce the heat energy back into the house instead of letting it through to the outside. The coating is applied to the glass by a process called magnetron sputtering. The glass is passed through a high-vacuum chamber and exposed to a mist of electrons given off by such metals as tin, indium, silver, copper or titanium. The mist condenses when it hits the glass, forming an invisible layer that is then covered with a similarly invisible coating of silver to prevent the oxides from rusting upon exposure to air.

The job of the oxide coating is to discriminate between different types of radiation; it lets through certain wavelengths and repels others. Low-E-coated glass in a double-glazed sealed unit increases the window's R-value to 3.3 and reduces the heat loss by 35 percent.

No sooner had AFG come out with low-E glass than a second major

WINDOW ENERGY RATINGS

GAINS/LOSSES (watts/sq. m)	TYPICAL DOUBLE-GLAZED	HIGH-PERFORMANCE FIBREGLASS
SOLAR GAINS	+44.8	+31.8
HEAT TRANSFER	−72.2	−24.3
INFILTRATION	−3.0	−0.2
TOTAL	−30.4	+7.3

window improvement was introduced. Engineers understood that in a double-glazed unit, heat loss occurs by conduction through the inside glass, by convection across the airspace, then again by conduction through the outside glass. That's why they limited the airspace to about half an inch: the larger the airspace, the more air movement and therefore the more convection of heat from pane A to pane B. Low-E coatings cut the conductive process, and in 1983, manufacturers found a way to reduce convection: they filled the airspace in low-E windows with argon, a heavier-than-air gas that suppresses convection by slowing down movement. A sealed unit with two low-E-coated glazings and argon in the airspace has an R-value of 4.2.

If the glass itself is the rock, then the hard place is the spacers that keep the glazings apart and the frames that hold the sealed units. In most windows in the mid-1980s, these spacers – and often the frames – were made from aluminum. Even before the concern about aluminum's embodied energy, designers realized that aluminum frames conduct 3,000 times more heat than the glass within them. The switch to wood or fibreglass frames was easy. Not so with aluminum spacers, which have similar conductive properties that can often neutralize the advantages of low-E coatings and argon fillings. Experiments were made with wooden frames and vinyl spacers, but the industry standard is still aluminum. Some manufacturers have married vinyl or fibreglass with the aluminum in the spacers, but by far the majority of double- and triple-glazed units installed in new houses are spaced with aluminum.

The Dorwin windows have frames made by a Winnipeg company, Omniglass, out of shredded fibreglass mixed with a resin and produced by a process called pulltrusion, which means that the mixture is squeezed through a die with the desired shape into a long strip that is cut off at

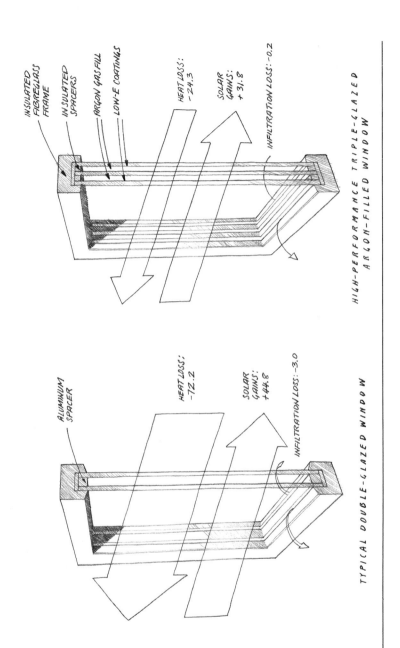

INSULATED FIBREGLASS FRAME

INSULATED SPACERS

ARGON GAS FILL

LOW-E COATINGS

HEAT LOSS: -24.3

SOLAR GAINS: +31.8

INFILTRATION LOSS: -0.2

HIGH-PERFORMANCE TRIPLE-GLAZED ARGON-FILLED WINDOW

ALUMINUM SPACER

HEAT LOSS: -72.2

SOLAR GAINS: +44.8

INFILTRATION LOSS: -3.0

TYPICAL DOUBLE-GLAZED WINDOW

Most windows represent a net heat loss, letting more heat out than they admit in. The Greenhome's net-gain windows are triple-glazed and argon-filled and have insulated spacers that do not act as thermal bridges; the frames are also made of insulated fibreglass rather than aluminum.

117

the desired length. Dorwin strengthens each corner with a right-angle brace and fills the frame with a non-CFC-blown polystyrene foam, then adds a triple-glazed, low-E-coated, argon-filled sealed unit (called an IG unit, for insulating glass) put together by Sunlite, a Toronto firm specializing in low-energy window technology. The use of Edgetech's Super-spacers—spacers that combine a small amount of aluminum for strength with a large amount of silicone for insulation—gives the Dorwin window its high energy rating. Edgetech's owner, Michael Glover, says Dorwin windows made with his Superspacers have a centre-window R-value of 8 and adds that if the 5 million square feet of windows installed in Canada each year were upgraded from R-2 (typical double-glazed aluminum windows) to R-8, then in 20 years, the net energy savings would be 100 petajoules per year, the equivalent of the annual output of Ontario Hydro's Darlington nuclear generating station. Put in more manageable terms, a homeowner with R-8 windows today would save more than $240 in space-heating bills every year for the life of the house.

□ □ □ □ □

At one of the earlier design meetings, held in March to discuss the question of insulation, the first thing agreed upon was that the Greenhome would not be insulated with fibreglass batts.

"It's deadly," Elizabeth White said flatly. "It comes with a skull and crossbones on the package."

The movement away from fibreglass insulation, though by no means universal in the housing industry—most builders and homeowners still "think pink"—began in October 1987, when the Housing Resource Centre and the American Council on Hazardous Materials hosted a conference in Cleveland, Ohio, entitled "Blueprint for a Healthy House." At the conference, a study was presented that linked glass-fibre insulation—fibreglass, rock wool and slag wool—with lung cancer. The study mentioned work done by the National Cancer Institute in the early 1970s, in which researcher Merle Stanton showed that fibres smaller in diameter than 1.5 microns and longer than 8 microns, when surgically implanted in rats' lungs, caused cancer. (Given that a micron is one-millionth of a metre, it must have been a delicate operation.) Earlier studies in Canada had linked minute, naturally occurring particles in asbestos to lung cancer; Stanton's work showed that what applied to asbestos also applied to man-made mineral fibres (MMMFs). His results had little impact on the scientific community, however, until 1986, when Philip Enterline and Gary March of the University of Pittsburgh showed that an examination of health records of 16,730 factory workers in 11 fibre-glass and 6 mineral-wool plants in the United States and 83,000 similar workers in Europe revealed a higher-than-average cancer rate. As a result, the International Agency for Research on Cancer designated all MMMFs as "possibly carcinogenic to humans." Fibreglass's earlier classification as a "nuisance dust" has been upgraded to more or less the same degree of seriousness as asbestos and cigarette smoke, hence the warning on the package.

It's true that, so far, heightened cancer levels have been found only in people who work in plants that produce mineral-fibre products. But as David Tovey, writing in *Harrowsmith* magazine in 1988, pointed out, "more than 5 million tons of MMMFs are produced in the world annually, and almost every house has some form of fibrous insulation in it. What are the rates of exposure for people who live in those houses and for contractors who install glass-fibre insulation day after day?"

"There's also the embodied-energy question," John added almost as an afterthought at the meeting. Fibreglass is to glass what candy floss is to sugar: it's made by heating glass to its melting point and spinning the molten stuff into thin fibrous shreds. Glass melts at somewhere around 2,000 degrees F, and the energy required to bring it up to that temperature is considerable, sending tons of carbon dioxide into the Earth's atmosphere and contributing to global warming. The same applies to Roxul, an insulation made of spun steel and sold in batts that resemble fibreglass batts, except that they are grey, stiffer and easier to handle. I've worked with both and much prefer Roxul, because its mineral fibres are coated with oil so that they do not float into the air and lungs; there is a lot less coughing and scratching after a day of poking the stuff into a house's innumerable nooks and crannies. Their R-values are about the same—up to R-15 for 4 inches of fibreglass, R-13 for the same thickness of Roxul. But from an embodied-energy standpoint, they were equally inappropriate for the Greenhome.

So mineral-fibre batts joined asbestos and urea formaldehyde foam insulation (UFFI) on the Greenhome's list of nonusable insulating materials. Styrofoam SM, the blue hardboard foam insulation made by Dow Chemical, was also added to the No list. The problem with Styrofoam is that during the manufacturing process, chemicals generally lumped together under the label chlorofluorocarbons, or CFCs—the correct chemical designations are CFC_{13}, commonly known as Freon 11, and CF_2C_{12}, or Freon 12—are released into the atmosphere. They rise to the Earth's stratosphere (15 to 30 miles up), where they are broken down by ultraviolet (UV) radiation from the sun. When broken down, CFCs release their chlorine atoms, which in turn break down ozone (O_3). To get a sense of the magnitude of the problem, consider that a single molecule of chlorine has the potential to destroy up to 100,000 molecules of ozone, and that a single extruded-polystyrene coffee cup made with CFCs can contain 1 billion molecules of CFCs. The ozone layer is supposed to absorb UV radiation; when it is destroyed, the UV rays penetrate through to Earth, where they do such unpleasant things as increase our risk of skin cancer and eye cataracts and contribute to global warming. Since the phenomenon was first detected in 1974, the ozone layer has undergone a general depletion of about 1 percent per year in Canada, with two major holes—no ozone at all—found over the south and north poles each spring.

Since the Montreal Protocol was signed in 1987—when 24 signatory nations pledged to reduce ozone-unfriendly chemicals (including halons, carbon dioxide, methane, nitrous oxide and ground-level ozone as well as CFCs) to 50 percent of 1986 levels by 1999—CFC production has declined sharply. CFCs take from 10 to 100 years to reach the ozone layer, however, so depletion will continue for at least the next century, and some say the hole over Antarctica will remain until current emission levels drop by 85 percent, or else forever, whichever comes first. E.I. Du Pont de Nemours & Co., the sole producer of Freon, has promised to surpass Montreal Protocol guidelines by completely phasing out CFCs by the turn of the century and has moved toward that goal by coming up with an alternative foaming agent—hydrochlorofluorocarbon, $HCFC_{22}$, or Form-a-Cell—that is 95 percent less depleting than F11 or F12. HCFC is used in aerosol spray cans but doesn't work for foam insulation because it escapes too quickly for the foam to retain its insulating properties. Since 1989, Dow Chemical has used a non-CFC- or HCFC-blown process to make its Styrofoam SM. Even so, says John, foam-insulation gases still contribute to the greenhouse effect, and he doesn't want to use this kind of insulation in the Greenhome. "Besides," he says, "none of them use recycled materials."

The only insulation that does use recycled materials is cellulose. More than 70 percent of all the trees logged in Canada each year are used by the pulp-and-paper industry, and until recently, almost all the paper products produced from those trees—newsprint, toilet paper, paper towels, coffee filters, and so on—ended up in landfill sites a few days after they were used. Cellulose insulation is made from recycled newspaper that is chewed up into a blowable mass, mixed with borax as a fire retardant and with a latex binder to keep it together, and sprayed into the wall cavities between the trusses (it has to be sprayed into a well-

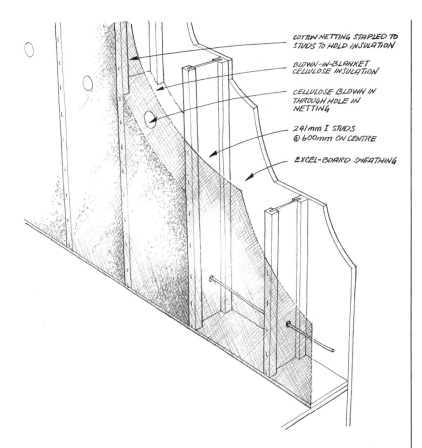

COTTON NETTING STAPLED TO STUDS TO HOLD INSULATION

BLOWN-IN-BLANKET CELLULOSE INSULATION

CELLULOSE BLOWN IN THROUGH HOLE IN NETTING

241mm I STUDS @ 600mm ON CENTRE

EXCEL-BOARD SHEATHING

Cellulose insulation is made of recycled newsprint mixed with a borax fire retardant and a latex binder that makes it adhere to the wall cavity, reducing shrinkage. Blown in through a nylon mesh, 9½ inches of cellulose gives an R-value of 37, nearly twice the industry standard.

defined space, like a box, rather than against a continuous wall). Dry cellulose insulation isn't exactly new: Saskatchewan Conservation House used it in the floor and ceiling (and fibreglass batts in the walls), but because there was no latex in it, it could only be used in horizontal applications. The new generation of wet-blown cellulose insulation, which has latex in it, adheres to wood with almost no sinking or shrinking. A similar product, with cornstarch instead of latex, was used in the Brampton House, and Elizabeth noted at one of the meetings that it has a moisture content of about 14 percent and needs about 36 hours to dry.

Cellulose is installed by a system known as the blow-in blanket system, or BIBS. A small-opening nylon mesh, called an Insulnet, is stapled to the wood frame, almost like a polyethylene vapour barrier. The cellulose is blown in through holes punctured in the mesh until the entire cavity is filled. The 9½ inches of cellulose gives an R-value of 37 and has an air-infiltration rating of only 2.3 cubic feet per minute, compared with 7.2 cubic feet per minute for standard fibreglass batts. Cellulose is also the insulation used in the Nova Scotia Envirohome. Dale Eastman, the Envirohome's project manager, says that when his framer came in one morning after the heaters had been on all night, the house was so warm, he turned the heaters off. He worked all day with them off, and when he went home that night, he forgot to turn them on again. He came in the next morning and worked until 4 o'clock in the afternoon before he realized that the heaters were still off.

The material used for the exterior sheathing against which the cellulose is sprayed is called Excel Board. Manufactured by the American Excelsior Company – an employee-owned firm based in Arlington, Texas – it is a mixture of polyisocyanurate foam and aspen fibres that comes in ¾-inch 4-by-8-foot panels, each of which weighs 20 pounds (less than half the weight of chipboard) and has an R-value of 3. Excel Board is nailed to the exterior of the house over the wall trusses. Later it will be strapped horizontally for the siding. The foam board is cream-coloured, with the wood fibres embedded in it, so looking down at a piece of Excel Board is like looking into a bowl of milk and shredded wheat.

Before the cellulose is blown in, while the wall cavities are still open, the plumbers and electricians install the Greenhome's wiring and plumbing – the house's guts and nervous system. Here is where cellulose's practical advantage comes into play: with mineral-fibre batts, the insulation has to be tucked and prodded around wire conduits and vent stacks, and no amount of care will result in an absolutely unbroken thermal barrier at places where such hardware breaks through the surface of the wall. Cellulose can be blown in to fill those holes.

After the cellulose, all that remains to be done to complete the shell is to install the air/vapour barrier (AVB). This is a fairly simple process that has an extremely complicated theory and history.

In a traditionally built house, something like 60 percent of the energy requirement is needed to replace heat sucked out by air leakage through the wall. It didn't take long for low-energy builders to realize that the best way to reduce air leakage was to line the inside of the house with a layer of polyethylene vapour barrier. The effectiveness of the vapour barrier was championed by Harold Orr and demonstrated in the Saskatchewan Conservation House in 1977; it was picked up and taken to its extreme by Jack Diamond in his Flair Home project in Winnipeg. It was incorporated into the EMR's R-2000 program and is now part of the National Building Code. No low-energy builder now seriously challenges the need for an effective airtight barrier in an energy-efficient house. There has been, however, a fair amount of disagreement about the method by which the most effective, most airtight air/vapour barrier is to be achieved.

Now for the theory. All one really needs to know about the philosophy behind vapour barriers is that warm air can add more moisture than cold air and that when warm air cools down, it deposits its moisture on the nearest object to hand. When that object is inside your wall, the moisture content can build up until it is more accurately described as water content, and wooden walls with high water content are great places for fungi to grow, and fungi are the sole cause of wood rot: wet walls rot. In old, drafty houses in which cold air whistles into the house through openings in the wall, moisture buildup is not a problem. In modern, highly insulated houses in which the inside warm, moist air is more likely to be whistling *out* through openings in the wall, the warm air will deposit its moisture inside the wall cavity, and as a result, shortly after you begin to notice water dripping from your ceiling fixtures, your house will begin to rot out from under you. The solution: stop that warm air from whistling out through the walls.

The first barriers – tar-impregnated paper with foil backings – were not very airtight. They were followed by sheets of polyethylene stapled to the inside of the exterior wall studs and ceiling joists. No caulking was used to seal the joins, because at the time, no known caulking adhered well to plastic. Such caulking as existed was designed either for exterior use – filling in holes around window frames to keep water out, and so on – or for such interior uses as grouting between bathroom tiles. The problem with this system, if it could be called a system, was that warm air continued to whistle out between the sheets, the sheets themselves pulled away from the staples, and there were huge air passages left where such things as electrical wiring and outlet boxes broke through the drywall.

The answer was to tighten up. The first breakthrough came when it was discovered that acoustical sealant adhered well to polyethylene. This caulking could be used to fuse sheets of plastic together to form a continuous air/vapour barrier. Joins between sheets were caulked from ceiling to floor, first by overlapping two sheets the distance of one stud bay, then by folding the edges of each sheet, caulking the fold and stapling through the caulking. Some particularly attentive builders cut little squares of cardboard and stapled through them to prevent damage to the plastic. After a while, it became apparent that tiny holes in the plastic itself were not the main culprits in air leakage; more attention needed to be paid to the gaps around window and door openings, which accounted for nearly all the air exfiltration from a house's interior.

In the Greenhome, before the window and door frames were installed, a 24-inch-wide strip of 6-mil poly was wrapped around the inside of each rough opening, and this was caulked to the vapour barrier on the wall so that any air leaking in between the frame and the wall would be directed out through the wall cavity rather than into the interior of the house. Between the basement wall and the upstairs wall, where so much air leakage occurs normally, some builders recommend routing the AVB from the basement wall out through the top of the concrete foundation wall, up the outside of the floor joists, then back into the interior of the upstairs house wall so that there is a continuous AVB from basement floor to upstairs ceiling. This was not necessary in the Greenhome, because the exterior Excel Board sheathing (milk and shredded wheat), which extends well below the sill, with all seams taped and tied to the Baseclad that goes down to the footings, is moisture-resistant enough to do the same job.

All the seams in the AVB were folded, caulked and stapled through the caulking. The caulking used was DyMonic sealant, a modified polyurethane joint caulking whose main advantage over other types of sealant is that it comes in 30-mL tubes—huge sausagelike rolls six times bigger than normal tubes, thus producing one-sixth the waste.

As Werner walks around the site picking up scraps of insulation, which he is going to throw into the attic above the garage rather than haul to a landfill, he tells me how other houses are insulated. "Normally," he says, "the contractor subcontracts the job out to a person who installs the fibreglass batts, tacks on the vapour barrier, does the cleanup and gets paid a grand total of $10 per bale of insulation. Now an average house uses maybe eight bales of insulation, so the guy gets paid less than $100, including cleanup. You can imagine how much care and attention to detail you get in the vapour barrier for that kind of money," he says, shaking his head.

"We sure do things differently here," he adds.

THE MECHANICALS

A HOUSE IS A MACHINE FOR LIVING IN.
— LE CORBUSIER

BY CHANGING THE HOUSE, WE INEVITABLY CHANGE SOCIETY.
THIS IMPLIES A SHIFT IN DOMESTIC ARCHITECTURE FROM
THE LE CORBUSIEN PARADIGM OF THE MODERN HOUSE AS A
"MACHINE FOR LIVING" — A DIRECT EXTENSION OF THE
FOSSIL-FUEL-POWERED TECHNOLOGICAL SOCIETY — TO THE
CONCEPT OF ARCHITECTURE AS AN EXTENSION OF THE
NATURAL ENVIRONMENT, FUELLED BY THE SAME FORCES
THAT DRIVE THE REST OF THE BIOSPHERE.
— SEAN WELLESLEY-MILLER, *TOWARD A SYMBIOTIC*
ARCHITECTURE

On December 2, the day Maurice Strong became chairman of Ontario Hydro, I was in the basement of the Greenhome watching three technicians from Union Gas install the natural-gas lines. The newspapers that morning had been full of bad news from the electricity company. It was $33.3 billion in debt, an amount almost exactly equal to the national deficit, which was also announced that day. The utility had had to increase its rates by 11.5 percent in 1992 and was promising to increase them by another 8.5 percent in 1993. In addition to raising rates, said its incoming chairman, whose $425,000 salary had been criticized in some quarters, Hydro was going to have to cut expenditures. Among the cuts was the $2.3 billion it had planned to spend on reducing emissions from its coal-powered generating stations. "I see our goal as that of trying to ensure that Ontario is amongst the world's most energy-effective and competitive economies," said Strong.

"He's the captain of a ship that's already hit the iceberg," commented Tom Adams, a researcher with the environmentalist group Energy Probe. "His job is to save the passengers." It's a ship with fewer and fewer passengers (demand for electricity has declined 3 percent a year for the past two years), a skeleton crew (one out of every five of Ontario Hydro's 28,500 employees is being asked to take early retirement) and diminishing prospects for the future (one of the Union Gas installers at the Greenhome, Terry Eckersberg, told me that morning that 97 percent of all new houses in the Waterloo area are heated with gas instead of electricity). He also said that Union Gas was averaging 12 conversions a day as owners of older houses switch to natural-gas furnaces. "The conversion we're most proud of," he said, grinning, "is the Waterloo Hydro building. We did that one last fall. It's pretty interesting when Hydro can't afford to heat its own buildings with electricity, don't you think?"

Terry and the other two installers, Larry Noble and Greg Thompson, bantered as they ran the main gas line from the meter beside the garage to the mechanical room in the basement. On the day Maurice Strong took over Ontario Hydro, morale was high at Union Gas.

The line they were installing seemed specifically designed to raise the morale of installation teams. Instead of the rigid lengths of copper or black steel that usually carry natural gas into and throughout a house, they were using a brand-new product called Wardflex, made by Hitachi in Blossburg, Pennsylvania. Ribbed stainless-steel tubing coated with yellow plastic, it looked a lot like those long, flexible necks on 5-gallon gasoline cans. The ribs allow it to bend around corners like rubber, dispensing with the soldered elbows, pots of flux and acetylene blowtorches required for copper tubing. When it needs to be attached to a fitting, an inch or so of plastic is peeled back to expose 10 rings of the tubing, the last few rings are shoved into a copper flange, and the flange is tightened with a wrench, compressing the rings to make an airtight seal. This was only the second time Wardflex had been used in Canada—the first was in an apartment complex near Leamington, Ontario—and it didn't even have a Canadian distributor yet. Union Gas was treating the Greenhome installation as a sort of on-the-job training. The gas company was so keen on the product that it was donating the tubing and the labour free of charge.

"It's great stuff to work with," said Larry. Until this job, he'd been working in the customer-service department, and he was very happy to be doing installations, especially with a new product. "We've been doing about half our installations in copper and half in steel; copper we can do in a few hours, but steel can take up to two days. This new stuff is faster than copper even." The ⅝-inch main line from the outdoor meter is connected to a specially designed manifold in the mechanical room, from which no fewer than six branch lines fan out to the Greenhome's various gas-burning appliances: the stove and clothes dryer in the kitchen; the fireplace in the great room; the barbecue on the upstairs deck; and the furnace and hot-water heater in the basement. Each branch has its own shutoff valve at the manifold. When the installers had the lines in, the flanges tightened and the ends of the tubing capped, they could test the system to ensure that it was leakproof. They did this by filling the lines with air: Greg went outside to the meter, attached an ordinary bicycle pump to a nipple, then pumped like mad to bring the whole system up to 50 pounds of pressure – "According to Ontario Regulation 244-89, Section 8.25.2," he said between pumps, "we're only required to bring it up to 25 psi and hold it there for an hour. When the gas is in there, it's only at about 4 psi, but we test it at 50 just to be on the safe side."

When the needle on the meter hit 50, we all congratulated Greg and trooped back to the mechanical room to see whether the seals were holding. Terry sprayed the fittings from a bottle filled with a soap-and-water solution and frowned as dozens of tiny air bubbles frothed up around the rim of one of the shutoff valves: a leak.

"Maybe copper is better after all," said Larry.

"No problem," said Terry, tightening the flange with his wrench. The bubbling increased. "Hmmm." He loosened the flange until all the air in the line escaped with a hiss of frustration and pulled out the tubing; the end of it, jagged and compressed into the flange, had punctured the neoprene gasket in the shutoff valve, causing the leakage. Terry trimmed the sharp points off the tubing with a file, replaced the gasket and reconnected the tubing to the valve. "Okay, Greg, crank 'er up again, and don't spare the horses."

This time the valve held air, and the Union Gas crew calmly set about undoing all the flanges in the house, trimming the burrs off the ends of the tubing, replacing the gaskets and retightening the connections. In some cases, they cut off the whole end of the tube and started fresh, counting back an extra ring to get more compression in the flange. "Next time," said Larry, making mental notes for some future installers' manual, "I think we'll cut the tubing with a hacksaw instead of a tube cutter. We'll get a cleaner cut. And we'll recommend compressing back four rings instead of three. That should guarantee a tighter connection."

An hour later, the needle on the meter was still holding firm at 50 psi, and the crew, satisfied that the lines were tight and ready to be hooked up to the appliances, whenever they might appear, piled into their blue propane-powered Union Gas van and drove off for lunch.

□ □

More light!
—Goethe's last words

Ontario Hydro may think of itself as the victim of the recession, but its current difficulties are more the result of shortsightedness. The problem is not that it sells too little electricity but that it produces too much.

According to its own estimates, it has been producing a surplus of megawatts since 1991 and will continue to do so until its 3.7 million customers catch up with the utility's electrical supply again in the year 2009.

Its recent marketing strategy has been consistent with its way of thinking: after a decade of cajoling customers to use less electricity, it is now subtly stroking them back into using more. The tactic is to convince us, subliminally if necessary, that electricity is energy-efficient. In a recent brochure entitled "Best on the Block: The R-2000 Home" prepared not by EMR, which sponsors the R-2000 program, but by the Canadian Electrical Association — a Montreal-based organization whose mandate is "to represent Canada's electric utility industry" — the "R-2000 home" of the title mysteriously becomes "the R-2000 all-electric home" in the text. "R-2000 all-electric homes are built with top-quality materials by specially trained builders," we learn in the introduction. "Electric heating systems are properly sized to provide even heating and improved comfort and economy. Modern appliances and lighting add style and convenience," and so on and so on. The message is simple but misleading: all R-2000 houses, it suggests, are electricity-dependent. "Electricity," the brochure says, "is the logical choice of energy in the R-2000 home." And later: "Electricity offers the perfect conversion of energy to heat." And in a final, spectacular burst of enthusiasm: "Electric heating is 100 percent efficient (or higher with a heat pump)." These statements sound good; unfortunately, none of them is true. The fact is, as Lawrence Solomon notes in *The Conserver Solution*, "it's inefficient to use electricity for low-grade energy needs such as home heating. It will do the job, but then caviar could substitute for rock salt in de-icing your driveway."

Because of wastage and transmission-line resistance, only a small fraction of the electricity produced at the generating plants is ever delivered to our houses, R-2000 or not. According to Phil Elder, the author of *Soft Is Hard*, nuclear generating plants create three times more heat than is necessary to produce the electricity they do. The waste heat is discharged as steam through those beaker-shaped stacks that have become the ubiquitous symbols of the nuclear age. Besides the visible waste at the generating station, running electricity along transmission lines drains even more energy (natural gas, by the way, also suffers some pipeline loss, about 10 percent between Alberta and Ontario). Some studies put the overall efficiency of an electrical generating station at around 30 percent, but even at 50 percent, a generating station would be wasting as much energy as it yields. Saying that electric heating is 100 percent efficient is like saying that a Cadillac is 100 percent efficient because it burns up all the gasoline pumped into it.

When Ontario Hydro sent a representative to Tim Mayo's first Advanced Houses information meeting, it did so because it was interested in supporting the idea of electricity conservation, definitely not in promoting alternatives to electricity. When the Greenhome design team decided that the space-heating and cooking needs of the house would be fuelled by natural gas, Ontario Hydro "sort of lost interest," says Steve. "They wanted us to build a Smart House, and we didn't think a Smart House was all that smart."

The Smart House is not so much a house as a concept, part of a global push toward turning houses into miniature versions of the starship *Enterprise*. A house is "smart" to the extent that its electrical system, at least in theory, is controlled by a central computer which runs everything in the house that uses electricity; that means just about everything but the kitchen sink.

Traditional houses may be wired with half a dozen separate and distinct wiring systems—two or three for lights, others for the stove, the furnace, the other appliances, the smoke detectors, et cetera. Each of these systems has its own set of plugs, which are often not interchangeable: you can't plug an electric stove into a lamp outlet, for example, or a stereo into your telephone jack. In a Smart House, you could. Smart House designers came up with a single, thick-as-your-wrist hybrid cable marketed under the name Smart-Redi wiring, made up of an army of separately shielded but essentially interconnected wires, something like a telephone cable. The cable is connected to all the plugs and switches in the house, so anything from a telephone to a furnace can be plugged in anywhere. All these lines run to the central computer, which monitors and controls the appliances that are connected to them.

This is possible because everything that can be plugged in contains a microchip which sends and receives signals from the central computerized power controller: "Hello, I'm a clothes washer, and I need 1.05 kilowatts of power for the next 14 minutes," and the controller releases the requisite amount of electricity. No chip, no power. If your 18-month-old decides to fix the toaster oven with a fork, nothing unimaginably horrible will ensue. If an appliance shorts out or otherwise ceases to function, the controller will simply withdraw power from it, print a message on your television screen—"Your kettle just blew up. Don't worry, I've taken care of it"—alert the house's security alarm system and even, if the problem is serious enough, call the fire station (or you at work) with a voice message. The controller can be programmed to, say, turn up the furnace thermostat an hour before you leave work, or lock the table saw so that the kids can't play with it when they come home from school, or turn on lights in a cunningly devised pattern while you're away on vacation, or open the louvred shutters on the sunwing windows when the inside temperature reaches a predetermined level.

Introduced in the United States in 1984 by Smart House Development Venture Inc., the project is sponsored by the Research Foundation of the National Association of Home Builders and the Electronic Industries Association (through a group called CEBus, for Consumer Electronics Bus, a company formed to develop "smart" appliances), and at least initially, it had the support of most of the major utilities as well as such multinationals as General Electric, AT&T, Lennox, Shell and Du Pont. But the technology has not progressed as rapidly as its originators had hoped. The plan was to have 5,000 Smart Houses up and running by 1987; this was downscaled to one, in Upper Marlboro, Maryland, by 1990. There are a few semi-Smart Houses in the United States and Canada, but the revolution is still waiting in the wings. One of the problems is that CEBus hasn't come up with enough "smart" appliances to accommodate a modern house, and if only a few appliances can be controlled by computer, the concept breaks down. Another problem is that so far, mainstream home buyers are a lot less interested in living in Smart Houses than electricity companies are in building them.

Smart Houses have been built in Japan and Holland. The TRON House in Tokyo is probably the ultimate house-as-machine; it is run by three interfaced personal computers linked to a microprocessing central control computer that operates everything from the house's three separate heating and cooling systems to each of its 127 windows. TRON stands for The Real-Time Operating System Nucleus and was designed by Ken Sakamura, a "computer architect" at the University of Tokyo. The house is fitted with its own rooftop weather station and dozens of microchipped sensors inside and outside that constantly send signals to the computers; for instance, instead of a thermostat that measures ambient air temperature in a room and turns on the furnace when it drops below a certain point, TRON's thermographic sensor system reads how many people are in a room, measures the infrared radiation emitted from their skin, records their surface body temperatures and adjusts the heating system accordingly — every five seconds.

The few North American versions that have been built are much less ambitious. There are two in Canada. The first, a 6,000-square-foot monster showcase called Manor Georgian, which Calgary Home Builders Association president John Klassen refers to as "a gorgeous revolutionary castle," opened in Calgary in May 1992, and a second one, north of Toronto, was announced in November of that year. Both have Smart-

Redi wiring and some smart appliances: the Calgary house, for example, is heated by natural gas, and its central communication system can signal the water heater to preheat water just before the clothes washer is scheduled to start up.

As Vic Sussman pointed out in a 1988 article, the Smart House concept is not aimed only at the technophile who loves to play with gadgets. Its professed goal is to save electricity: "Many utilities are interested in the Smart House project," he wrote, "because they believe it will lead to energy conservation. A house whose lights, heating and cooling are automatically regulated according to its inhabitants' occupancy patterns could use much less power than a standard house. And because Smart House controllers might ultimately download their information to a utility's database (say farewell to the noble meter reader), utilities could have access to far more sophisticated information on energy-consumption patterns."

Maybe so, but even within the electronics industry, interest is lukewarm, and outside the industry, support for the project has never waxed. "The idea worries me," Robert Sardinsky, an associate of Amory Lovins at the Rocky Mountain Institute in Colorado, told Sussman. Sardinsky believes that the modern house is "already totally dependent on support systems beyond its walls" and that the Smart House would simply move homeowners "one more layer away from control." People enjoy looking after their own homes, Sardinsky argues, "so perhaps we're depriving people of a human experience by automating. I think we may lose some of the spirit of home life by making houses self-tending."

Ironically, we may also lose some of the spirit of conservation. John Kokko thinks that if you remove the responsibility for energy conservation from the individual and give it to a computer or the utility, then the individual will eventually lose interest in it. "He or she will say, 'That's it, I don't have to think about it anymore. Someone else is taking care of it.'" During discussions about monitoring the Greenhome after it is sold, for example, when the options were to have the whole energy-monitoring system run from a single central meter or from a series of smaller meters, John argued for the latter. The homeowners, he said, have to be able to look at different meters to see whether the total energy consumption went down in a particular month because they lowered the thermostat at night or because they washed the dishes by hand instead of using the dishwasher. "I want to keep that hands-on experience," he said, "so the owners remain involved in the house, so the house doesn't become a kind of machine that runs along on its own."

When Ontario Hydro offered the Advanced Houses program a

$55,000 Honeywell TotalHome System—a somewhat downsized version of the Smart House—which adjusts a house's lighting, heating and cooling systems to fit the owners' living patterns and includes a complex security network that, according to the Canadian Automated Buildings Association, "allows the homeowner to create up to 16 patterns with combinations of armed and disarmed security points to fit their life styles," two of the houses accepted the offer: Innova House in Ottawa and the Neat House in Hamilton. One of the Honeywell features allowed the electrical utility to monitor the house's energy input and shut down unnecessary usages—hot-water heaters and clothes washers, for example, during peak periods—a sort of domestic demand-side management. The Greenhome team asked themselves the same question they asked about all the other aspects of the house: Is this the way we want houses of the future to go? In this case, the answer was no.

"Having a house in which your toaster can talk to your tea kettle might be fascinating for the first day and a half," says John, "but after that, I think I'd be worried about having Big Brother telling me when I can take a shower or cook dinner. Besides, we can accomplish all that without Ontario Hydro: we can set the manual timer on our dishwasher to start up after the 11 p.m. peak period, for example. We're using a lot of innovative electrical appliances that I should have thought Ontario Hydro would be interested in knowing about, but they're not. They just say we're not doing anything that Brampton House didn't do, which is quite simply not true. Brampton House was all-electric, for example, which, come to think of it, must be Hydro's point. They want us to say high-tech is the way of the future. We're trying to demonstrate how energy can be conserved every day. The two concepts appear to be mutually exclusive as far as Hydro is concerned."

Houses have been in the process of becoming high-tech machines for a long time, at least since the turn of the century, when appliances began to replace servants in North American middle-class homes. The first time electricity was used to run a machine was in 1889, when a New York grocer connected an electric motor to a coffee grinder. It wasn't long before electric motors were running everything from Singer sewing machines (1889) to washing machines (1909) and dishwashers (1918). "The character of the domestic home changed dramatically in the early 1900s," writes Witold Rybczynski in Home, "largely as a result of the availability of electricity."

The important point here is that in most cases, electricity was originally used not to heat things but to run small motors and to provide lighting. That is what electricity is good at. Take lighting, for instance.

Electric lighting has been around since 1882, when Thomas Edison built his first generating station in the Wall Street district of New York and began installing his new light bulbs in the surrounding square mile (and charging customers $5 per kilowatt-hour for using them). By the turn of the century, the price had gone down considerably and gasoliers were outmoded in every major city on the continent. Today, indoor and outdoor lighting accounts for a whopping 6 percent of all the energy – from oil, gas or electricity – consumed in North America. In 1978, General Electric estimated that lighting accounted for 24 percent of all electricity sold. That in itself was a lot of electricity – about 225 million kWh a year in Canada – and there is no reason to believe that the percentage has gone down by much, despite two decades of conservation programs. Only four years ago, physicist Arthur Rosenfeld calculated that the amount of electricity dedicated to indoor and outdoor lighting was still about 20 percent of the total North American output, with seven-eighths of that being used indoors.

By and large, that's too much. As I pointed out above, there is a vast amount of electricity to be saved by reducing the amount we waste; there is also a lot to save by reducing the amount we use.

Most of us, when we think about conserving electricity in the home, think of turning off lights when they're not needed or maybe switching to lower-wattage incandescent bulbs in the ceiling fixtures. And some savings can be realized that way. Replacing a 60-watt bulb in a table lamp with a 52-watt bulb will provide about the same amount of light and save about 15 kWh per year; make the switch on all 40 bulbs in an average house, and the result is 600 kWh a year, about $54 at current rates.

The Greenhome designers approached the problem on a more basic level, however. First they asked how much light is needed in a house. The amount of light is measured in footcandles, a unit defined by the Illumination Engineering Society (IES) as "the illumination produced on a surface all points of which are at a distance of 1 foot from a directionally uniform point source of 1 candela." In other words, 1 footcandle is the amount of light from a candle on an object that is 1 foot away from it or from two candles on an object that is 2 feet away from them, and so on. The amount of illumination you get from a light bulb is measured in footcandles per watt. When you shop for a light bulb, don't just look at the watts, which will tell you only how much electricity the bulb will use; ask about the footcandles, which will tell you how much light you'll get from it. If it's an imperially challenged shop, ask about lumens, the metric equivalent of footcandles. One lumen equals about 10 footcandles. This information is not printed on the bulb; if the shop clerk

134

RECOMMENDED LIGHT LEVELS

RANGES OF ILLUMINANCES

TYPE OF ACTIVITY	CATEGORY	Lux	Footcandles
GENERAL LIGHTING FOR			
Public spaces with dark surroundings	A	20-30-50	2-3-5
Simple orientation for short visits	B	50-75-100	5-7.5-10
Working spaces where visual tasks are only occasionally performed	C	100-150-200	10-15-20
Visual tasks of high contrast or large size	D	200-300-500	20-30-50
TASK LIGHTING FOR			
Visual tasks of medium contrast or small size	E	500-750-1000	50-75-100
Visual tasks of low contrast or small size	F	1000-1500-2000	100-150-200
COMBINATION GENERAL AND TASK LIGHTING FOR			
Visual tasks of low contrast and very small size over a prolonged period	G	2000-3000-5000	200-300-500
Extended and exacting visual tasks	H	5000-7500-10000	500-750-1000
Very special visual tasks of extremely low contrast and small size	I	10000-15000-20000	1000-1500-2000

AREA/ ACTIVITY	CATEGORY	AREA/ ACTIVITY	CATEGORY	AREA/ ACTIVITY	CATEGORY
General Lighting		Noncritical	D	Prolonged serious or critical	E
Conversation, relaxation and entertainment	B	Kitchen sink		Desk	
Passage areas	B	Difficult seeing	E	Primary task plane, casual	D
Specific visual tasks		Noncritical	D	Primary task plane, study	E
Dining	C	Laundry		Sewing	
Grooming		Preparation and tubs	D	Hand sewing	
Makeup and shaving	D	Washer and dryer	D	Dark fabrics, low contrast	F
Handcrafts and hobbies		Music study		Light to medium fabrics	E
Workbench hobbies		(piano or organ)		High contrast	D
Ordinary tasks	D	Simple scores	D	Machine sewing	
Difficult tasks	E	Advanced scores	E	Dark fabrics, low contrast	F
Critical tasks	F	Reading		Light to medium fabrics	E
Ironing	D	In a chair		High contrast	D
Kitchen duties		Books, magazines and newspapers	D		
Kitchen counter		Handwriting, reproductions and poor copies	E		
Critical seeing	E	In bed			
Noncritical	D	Normal	D		
Kitchen range					
Difficult seeing	E				

doesn't know how many lumens per watt a bulb will provide, call the customer-service department of your local utility or use the chart provided later in this chapter.

Studies conducted at San Jose State University in California in 1968 determined that for people with average eyesight, the amount of light needed to read a printed page comfortably is between 3 and 10 footcandles at the page or desk level (called the "task area"), and I believe it. In fact, I've checked it: I lit 10 candles, clustered them together on a table, held a book 1 foot away from the cluster and was able to read without the squintiest difficulty. Slightly higher levels might be needed for tasks demanding more discrimination or when there is less contrast in colour, such as in fine needlework or picking flyspecks out of pepper, but in general, if there are 10 footcandles of illumination on a task area, that task can be performed with reasonable efficiency. What's more, the study found that our mothers may have been wrong all the time: more light is not necessarily better for our eyes. In fact, higher levels of light can cause eye fatigue that actually decreases the eyes' efficiency.

Now, electrical engineers and architects, when determining how many light fixtures to place in new buildings, including people's houses, consult something called the *IES Lighting Handbook*. The IES is an august group of electrical engineers that regularly publishes its recommended lighting standards in a volume which has been called the bible of light by more than one architect. The IES, which is not unassociated with the lighting industry, conducts its own experiments, and its *Handbook* sets the standard light levels that are used by designers in the United States, Canada and England. Beginning in 1959, its recommendations began to climb, and as a result, building designers and governments around the world started boosting their light levels. In 1971, for example, the New York Board of Education raised its minimum requirements to 60 footcandles in classrooms, 100 in libraries and 150 on chalkboards.

These inflated levels persisted throughout the energy-stricken '70s and '80s. "We estimate that approximately 50 footcandles of light falling on a work surface is a necessary minimum if we wish to perform an easy task with comfort," states James L. Nuckolls in *Interior Lighting for Environmental Designers* (1976). To give an idea of just how much light that is, Nuckolls calculates that in "an average room measuring 25 feet by 17 feet . . . we would need over 135 incandescent 40-watt lamps to provide 50 footcandles of light."

In 1979, the Canadian government published *Energy Conservation*, a handbook for energy-conscious designers prepared by the Royal Architectural Institute of Canada. "Too much light is often provided in all areas of buildings," begins the section on lighting, which goes on to recommend that lighting levels in buildings be cut dramatically, in some cases to less than one-quarter of previous levels. The handbook stated that lighting levels between 11 and 44 footcandles were "sufficient for visual acuity and physiological needs," and noted that these levels were "a considerable reduction" from the 65 to 160 footcandles then provided in many buildings. The 160 footcandles referred to the amount of light the IES recommended for food preparation and cleaning areas in houses: did we really need the equivalent of a 1,250-candle chandelier hanging 5 feet above our kitchen counters for peeling onions or washing dishes? The Illumination Engineers said we did. Even the reduced level of 40 footcandles seems excessive, given Nuckolls' calculations.

With growing emphasis on energy conservation in recent years, the IES recommendations have gone back down somewhat: the 1992 *Handbook*, endorsed by Ontario Hydro, specifies 50 to 100 footcandles for kitchen counters and other task areas around the house, and that is what architects are designing to. Individual homeowners, however, can check light levels in their houses and adjust them accordingly.

Cutting down on lighting *levels* in the home is one way to reduce the amount of electricity consumed; another is to choose a type of lamp that delivers the most lumens per watt of electricity. (Note: architects and lighting engineers don't say "light bulbs" and "fixtures" the way the rest of us do; they say "lamps" and "luminaires.") The lamp most often used in residential luminaires is the incandescent lamp; in fact, the home is one of the last holdouts of the vacuum incandescent light bulb, which has remained virtually unchanged in design and composition since it was invented—simultaneously, by the way, in England by Sir Joseph Swan and in the United States by Thomas Edison—in 1881. Both men came to the conclusion that the way to get light from electricity was to run it through a partial vacuum along a scorched bamboo filament; as Moses observed during an earlier experiment, the wood in the glass bulb burned brightly but was consumed not, at least not for about 20 hours. In 1907, Edison substituted tungsten for bamboo, and we are still using tungsten filaments today.

Alas, the incandescent bulb is the least energy-efficient electric lamp on the market. A few examples will suffice to make this point. In 1939, a new kind of lamp was introduced in the United States, the fluorescent tube, which produced light not by burning a metallic element but by means of an electric arc that excites phosphors contained in the gas with which the tube is filled. In 1939, a 40-watt incandescent bulb gave off 20 lumens of light per watt; one of the new fluorescent 40-watt lamps, using the same amount of electricity, gave 40 lumens of light per watt. The next year, a 40-watt fluorescent was developed that produced 50 lumens per watt; and by 1965, 40-watt fluorescent lamps were producing 80 lumens per watt. After 1970, however, after OAPEC and the introduction of emergency energy-saving devices and techniques, and at about the same time that the IES decided we had to increase the light levels in our buildings by a factor of four, the lumen-per-watt count of both fluorescents and incandescents took a sudden and dramatic downswing: in 1976, a 40-watt fluorescent lamp gave off only 10 lumens per watt; and by then, fluorescent lamps were providing 75 percent of all the artificial light produced in the world.

Today, the most common fluorescent tube—the 4-foot 40-watt tube—delivers only around 50 lumens per watt. These tubes, however, are used almost exclusively in office buildings, schools and other more-or-less public or industrial settings. In the home, the standard is still the incandescent bulb, which is rated at about 13 lumens per watt. In 1984, 77 percent of the 6.5 billion lamps sold in the world were incandescents. Why do we keep using them? Probably because we think they give off a "warmer" light than fluorescent tubes do. Hundreds of studies have been conducted over the years to try to figure out what we mean by this, correlating light type, level, our expectations and efficiencies, and most of them have been inconclusive. We seem to prefer different kinds of light for different situations: our responses are affected by the colour of the walls in a room, the use to which the room is put, the amount of visual clutter, the need to discern colours or distinguish facial expressions, and a host of other factors that no architect can design for. Nuckolls reports, however, that no matter what we say we want, "in controlled conditions, a person can happily adapt to any one of the several apparently 'white' spectral blends." He also notes that in all the tests, one response seems to be consistent: "There is a preference for warmer tints of 'white' light in the presence of lower illumination levels." And we think this means we prefer incandescent lighting. As Canadian actress Gale Garnett wrote recently, "Any actor will tell you the human body looks awful under fluorescent light."

There are, however, lamps which give off warm white light that are not incandescent. High-pressure sodium lamps, for example, are rated "warm golden white," very close to incandescents, and they deliver up to 150 lumens per watt, depending on the wattage: a 35-watt high-pressure sodium lamp gives off about the same number of lumens as a 100-watt incandescent. The metal halide lamps used in many task-directed fixtures are rated "bright white light, good," and give 65 lumens per watt.

The newest lighting innovation in the low-energy market is the compact fluorescent bulb that the utilities have been encouraging consumers to use in recent months. Compact fluorescents are actually miniature fluorescent tubes often curled up in conventional-looking bulbs and fitted into magnetic 2-watt ballasts that can be screwed into sockets designed for incandescents (a ballast is a small transformer in the fixture that steps up the voltage high enough to activate the phosphors in the tubes). Like all fluorescents, they are more efficient than incandescents: they deliver about 50 lumens per watt, so an 18-watt compact fluorescent can replace a 75-watt incandescent. To get an idea of the energy savings made possible by compact fluorescents that can be fitted into incandescent sockets, consider the Exit signs we see in theatres, restaurants, hotel corridors and other public areas where the National Building Code requires them. Each one of those signs contains two 25-watt incandescent bulbs that burn 24 hours a day, which means each sign uses 438 kWh of electricity per year, at a cost of close to $40.00 (assuming the current rate of 9 cents per kWh). Both bulbs could be replaced by one 7-watt compact fluorescent that would cost only $5.50 a year to operate, for annual savings of $34.50 per sign. Now think about how many Exit signs there are in North America.

As if they needed another strike against them, incandescents don't fare very well on the actuarial charts either. Life expectancy for a standard 40-watt incandescent bulb is about 1,000 hours of use; a standard 40-watt fluorescent will last eight times longer, around 8,000 hours. A 35-watt high-pressure sodium lamp, however, will last 16,000 hours; a 50-watt high-pressure sodium lamp is rated at 24,000 hours. Metal halides and compact fluorescents are down around the 10,000-hour range, still 10 times longer-lived than incandescents. Life expectancy doesn't affect the direct energy efficiency of a bulb, but it does affect the embodied energy: it takes energy to manufacture a light bulb, and throwing away billions of nonrecyclable bulbs every year adds up: money for landfills is money out of all our pockets.

A recent study by the Electric Power Research Institute in the United States found that compact fluorescents are catching on, but slowly. The researchers phoned 328 users of compact fluorescents in Boston and San Francisco: 40 percent of them said that they would not use compact fluorescents again, that the price (around $10 each) was "outrageous." Still, 60 percent said they would.

Most of the lamps in the Greenhome are compact fluorescents recessed into ceiling "pot-light" fixtures with reflective interiors that direct more light into the living space, making it possible to use even lower-wattage lamps. In areas where lights are required for only short periods of time per use – closets, for example, or the fruit cellar in the basement – incandescents are all right, although lower-wattage bulbs are used (52 watts instead of 60, for instance). Other areas will be lit by fluorescent lamps, although not by the same tubes that are found in offices and stores. Industrial-strength fluorescent tubes are usually T12, which refers to their diameter: twelve-eighths of an inch, or 1½ inches. A typical T12 fixture with two 4-foot tubes uses 80 watts for the bulbs, plus another 17 watts for the ballast, and provides a total of 3,000 lumens, or about 30 lumens per watt. T8s are the newest fluorescents on the market. Narrower than T12s (1 inch in diameter), they use only 32 watts per tube and 2 watts per ballast and achieve up to 45 lumens per watt: a single 2-tube fixture with T8s in it gives off as much light as six 40-watt incandescent bulbs, while consuming about 28 percent of the electricity. Waterloo Hydro calculates that a house in Waterloo built to Ontario Building Code standards consumes 1,835 kWh of electricity per year for lighting; the Greenhome's annual lighting budget is just 500 kWh.

The Greenhome's outdoor lighting doesn't contribute much to that total, because most of it is solar-powered. The amount of energy wasted in North America in outdoor lighting amounts to nearly $1 billion per year. As Terence Dickinson pointed out in *Harrowsmith* magazine a few years ago, that is because most outdoor lighting fixtures send most of their light straight up into the nighttime sky as light pollution, rather than down onto the ground where it can do some good. He quotes David Crawford of the Kitt Peak National Observatory in Arizona, who estimates that 30 percent of all outdoor lights "do nothing useful" and adds that "in the United States, wasted energy from light pollution equals the entire electricity production of Ireland."

The Greenhome's three porch lights and driveway floodlight are all 17-watt compact fluorescents with magnetic ballasts (there may be some momentary flickering while the ballasts boost up the tubes, but electronic ballasts won't work at all in low temperatures), and its 10 yard lights, made in California by Seimens, collect solar energy during the day, store it in storage batteries and use it during the night to light up the walkway and patio areas around the house. Reflectors on top of the fixtures direct the light down onto the ground and can be either turned on and off manually or else left on to shine until the battery runs out of juice, which in most locations can take up to six hours. Either way, they use up no purchased electricity, do not add to the city's nighttime light pollution and don't require underground wires. They can even be picked up and moved around as needed. As Richard Reichard said at the design meeting when John brought one in to show the team, "Why doesn't everyone use these things?"

□ □ □

Le Corbusier's dictum that a house is "a machine for living in" has had no greater application than in the kitchen. As Witold Rybczynski has observed, it is in the area of "modern conveniences" that most of the changes in domestic arrangements have been made since the turn of the century. In 1912, Christine Fredericks wrote a series of articles for *Ladies' Home Journal* entitled "The New Housekeeping," in which she advocated applying to the home the same kind of efficiency concepts that were changing American factory layouts: storing tools close to where they are used, organizing kitchens to reduce movement between food-storage and preparation areas, and so on. She also established a test kitchen in her Long Island house in which she assessed new appliances for their efficiency. A few years later, Lillian Gilbreth, wife of efficiency engineer Frank Gilbreth, was hired by appliance manufacturers to test and endorse their products. Each new apparatus sent to her was carefully scrutinized for its ability to save the amount of work that actually had to be done in the real world of the kitchen rather than in the rarefied atmosphere of the lab or the salesroom.

Early advertisements for electrically powered appliances extolled the virtues of electricity, not because it saved time but because it did the work for us: a Victorian ad for an electric Blickensderfer typewriter, for example, showed two female typists, one looking harried and pounding away at an old manual contraption, the other smilingly tapping away at a brand-new Blickensderfer. "The old standard of efficiency: Work! Work! Work!" reads the caption under the first typist; and under the second: "The new standard of efficiency: Touch the key, Electricity does the work." It was not the speed of electric typewriters that was being promoted; it was the ease with which they could be operated. Similarly, an ad in Eaton's 1922 catalogue shows the Acme B electric clothes washer, identical to the hand-operated Acme C model except that it had a 6-horsepower electric motor attached to the crank that operated the agitator: "The hard work of rubbing, scrubbing and wringing clothes vanishes the moment you install this electric washer and wringer," reads

141

the ad copy. No mention of the Acme B getting the job done faster than Acme C, just that it "eliminates wash-day drudgery." There was also an Acme A, by the way, very similar to the electric model except that its agitator could be either hand-powered or attached to a windmill; it cost $16 compared with $85 for the electric model. Lillian Gilbreth's criterion was that a new appliance must reduce the amount of work a woman had to do, not increase the amount of work a woman could do. This is an important distinction. We don't seem to make it anymore.

Christine Fredericks, Lillian Gilbreth and even Timothy Eaton were concerned about making the home a more comfortable place in which to work. The goal of efficiency was to create more leisure: some ads actually showed housemaids resting beside an electric washing machine while the motor whirred comfortingly away. Women of the time did not, for the most part, work outside the home. In the past few decades, however, not only have many more women joined the paid work force but there has also been a huge increase in the number of single-parent families. As a result, there is less time for housework. Modern appliances are now sold as time-saving rather than labour-saving devices. New electric toasters toast faster than old toasters; they don't make us work any less. New irons heat up and start steaming faster; they don't make ironing any less work. I recently toured the small-appliances section of a local department store and confirmed most of my prejudices: a Hotstuff curling iron, for example, "heats up in 90 seconds!" and costs $8.99, but a Philips curling iron has a "30-second heat-up" and sells for $19.99. This tremendous time saver not only costs more to buy but also costs more to operate: the Hotstuff works on 8 watts, the Philips on 14. Once heated up, however, the Philips is just as much work for the user as the slower model.

This new emphasis on saving time has resulted in a huge increase in the number of electrical appliances available, with a corresponding increase in the amount of electricity devoted to operating them. The appliance section in the same department store carried a total of 43 different types of small appliances, from electric kettles to electric toothbrushes to fondue pots to rice cookers to cream whippers to electric fingernail-polish buffers to (I'm not kidding) electric nose-hair clippers. Very few of these items actually save us any significant amount of work – some of them even create work – but together they add up to an enormous amount of energy usage. The proportion of household electricity we use to operate our appliances has skyrocketed in recent years. In 1960, an average house monitored in Washington, D.C., used a total of 3,300 kWh of electricity a year for everything; the same house, in 1972, used 8,500 kWh. Today, according to a pamphlet published by Ontario Hydro, an average house in Ontario uses from 10,000 kWh to 30,000 kWh of electricity a year *for appliances alone*, despite the fact that family size has declined from 4.4 to 2.8 persons since 1972.

The Waterloo Greenhome has cut the amount of electricity used by its appliances down to about 3,000 kWh per year. The only difference the eventual owners will notice, however, will be in their electricity bills. They'll have all the major appliances—refrigerator, freezer, dishwasher, clothes washer and dryer—and quite a few of the minor ones: a coffeemaker, an electric kettle, a microwave oven. The savings will come from a more considered use and a more careful selection of the appliances.

In general, the electrical consumption of appliances has been declining for the past two decades. For instance, in 1978, the average two-door frost-free refrigerator/freezer used 2,000 kWh of electricity a year; by 1983, the same-sized unit used only 1,500 kWh. By 1992, it was down to just under 1,000 kWh. Within that average, there is a remarkable difference in energy performance—one of Maytag's models uses 780 kWh per year, while a Beaumark consumes 960 kWh per year—but in general, wattage used to run household appliances is half what it was 15 years ago.

Much of the credit for that goes to the Canadian EnerGuide program and its American counterpart, EnergyGuide, two government watchdog agencies that gather, collate and publish energy-consumption ratings for all major appliances sold in North America. The American program has been more successful than the Canadian version, it must be said: although the Canadian EnerGuide was put in place first, as early as 1977, the American government has given its energy watchdog more bite every year, while Canadian politicians have done nothing but pull EnerGuide's teeth. Originally, the Canadian program required all manufacturers to label their appliances with EnerGuide stickers stating the amount of energy the appliances consumed per month. Manufacturers did their own testing, but EnerGuide spot-tested the results and published its own directory. Most of EnerGuide's funding was withdrawn in 1986, however, and now the program merely accepts manufacturers' energy ratings and lists them in the directory; the Canadian government has passed no legislation requiring manufacturers to reduce the energy their appliances consume. In 1992, however, the United States passed a law by which all appliances manufactured there in 1993 could consume only 70 percent of the energy used by those sold in 1992. No such law has even been conceivable in Canada; as Brian Kelly, an energy specialist in Ottawa, put it, "standards are anathema to our federal Tories."

As a result, energy-consumption standards for American products are significantly stricter than ours, so much so that Art Rosenfeld, direc-

FUEL CONSUMPTION

ITEM	KWH/YR TYPICAL WATERLOO HOUSE	KWH/YR GREENHOME
GAS		
Space Heating	18,554	5,250
Hot Water	12,422	977
Clothes Dryer	1,680	546
Range	960	1,040
ELECTRIC		
Refrigerator	1,500	640
Freezer	1,200	
Dishwasher	720	252
Clothes Washer	1,374	144
TV/Small Appliances	1,086	546
Lighting		
Indoor	1,369	479
Outdoor	466	20
Fans	2,037	1,619
TOTAL	43,368	11,513
TOTAL COST @ $ .09/kWh	$3,903.12	$1,036.17

tor of the Center for Building Science at the Lawrence Berkeley Laboratory in California, calculated six years ago that if all appliances in Ontario households conformed to U.S. instead of Canadian standards, the energy saved would equal the output of one nuclear reactor. Today, the difference is even greater.

For the Greenhome designers, getting the energy consumption down to 3,000 kWh a year was not as easy as opening up the EnerGuide directory and ordering the most efficient appliances in it. Appliances for the Greenhome had to fulfil a number of requirements arising from environmental considerations, not just energy consumption: embodied energy, for example, and no CFCs. And even before those elements were factored in, the designers had to reconsider some of our basic assumptions about appliances. Take refrigerators: what *size* refrigerator does today's home require? Just as our house sizes have increased over the past 50 years despite declining family sizes, so has the volume of our refrigerators. According to *Consumer Reports*, an independent consumer-advocacy magazine that tests and rates products according to a wide variety of factors—sometimes even energy efficiency—the average refrigerator sold in North America today has 18 cubic feet of storage space. That is huge. Five years ago, the average was 15 cubic feet. Our house has had five people living in it, three of them teenagers, and our 13-cubic-foot Wood's has proven perfectly adequate, with plenty of space to store leftover food until it was rotten enough to transfer to the composter.

Ours is what is called in the trade an "all-fridge": it doesn't have one of those little freezer units mounted in the top of it. Those in-refrigerator freezers gobble up an enormous amount of energy and are completely unnecessary if there is a full-sized freezer somewhere else in the house. It may be inconvenient to have to walk all the way to the freezer for an ice-cube tray; but the savings to you and to the environment are significant (besides, most all-fridge models now have small compartments in them for freezing and storing ice cubes). The 1991 EnerGuide directory listed only five refrigerators without interior freezers, and their average consumption was 560 kWh per year (the largest of them, a 20-cubic-foot Sub-Zero model, used 588 kWh a year). The 1992 directory lists 16 all-fridges, still with an average annual energy consumption of 560 kWh. Are manufacturers catching on?

Refrigerator/freezers use more energy than all-fridges because they use a single compressor to cool both sections. In fact, in most North American models, the fresh-food section is cooled by cold air vented out of the freezer unit. This means that every time you open the refrigerator, you're also opening the freezer. The Danish manufacturer Vestfrost makes a two-door refrigerator/freezer with separate compressors for the freezer and fresh-food sections. The combined energy consumption of the two compartments is only 320 kWh per year, less than the most energy-efficient all-fridge on the North American market. Two Vestfrost refrigerator/freezers, with a combined capacity of 14 cubic feet of fresh-food space and 7 cubic feet of freezer space, require a total of 640 kWh/year. The most efficient North American 14-cubic-foot all-fridge combined with the most efficient North American 7-cubic-foot freezer would require 1,050 kWh/year. The Greenhome designers decided to go with the Vestfrosts.

The lower energy consumption of the Vestfrost refrigerators was, however, only one of their advantages. The real challenge in choosing a refrigerator these days is finding one that does not contain CFCs, either as a coolant or as a blowing agent in the insulation. The Vestfrost re- frigerator/freezers use HFC-134A as a coolant, and the foam insulation in their walls is made with the non-CFC blowing agent HCFC 123. Both are improvements over CFCs because the additional hydrogen atom (the H in the formulas) increases the weight of the gas, so when it is released into the atmosphere, it does not rise as quickly, and 98 percent of it is destroyed before it reaches the ozone layer. "It's not perfect," says John, still concerned about that escaping 2 percent, "but it's the best we can do for now."

Refrigerators haven't always been considered essential home appli- ances, and many of the products we keep in them even now don't need constant refrigeration. A tour of our own refrigerator yields several such items: eggs, apples, lemons, jars of pickles, a can of maple syrup, which are kept refrigerated either out of habit or because we bought too much too long before we could use them. Rational shopping, in other words, can be an energy-conserving activity: we spend a lot of energy and money refrigerating packaging.

Another household appliance that some might find inessential is the dishwasher. In the past, certainly, dishwashers used a lot more hot wa- ter than hand-washing did, and those who wanted to save energy by turning down the thermostat on their water heater to a reasonable tem- perature could not set it below 140 degrees F if they had a dishwasher. Today, not all dishwashers use the same amount of hot water. Although the average is 10.5 gallons per wash, individual models in the EnerGuide directory range from 8.7 to 12.5 gallons, depending on the size of the machine. This is still less water than is normally used in hand-washing dishes; a study conducted at Ohio State University a few years ago found that washing dishes by hand using the two-sink method required 13 gal- lons of water. However, the water used in hand-washing can be cooler than water required by a dishwasher – ours is down at 120 degrees, the lowest setting on our tank, and it's still so hot we have to add cold water to it when we wash the dishes. The average energy consumption for dish- washers in the EnerGuide directory was 950 kWh a year, including the energy needed to heat the water; it is based on 34 washings per month at 10 gallons of hot water per wash. Making sure that the dishwasher is always full when it is used, always soaking the dishes beforehand so that they only need to be washed once, and not using the "heat dry" option (which uses more energy and dries the dishes only slightly faster) can reduce water use – hence energy consumption – considerably.

The Greenhome's Miele dishwasher requires 640 kWh a year and uses only 4.7 gallons of water per wash. Miele is one of Germany's oldest and most reliable appliance companies; founded in 1898, it introduced the first electric dishwasher in Europe in 1929—it was a round, wooden top-loading tub with a pump-operated spray arm at the bottom, "very primitive and inefficient," says Miele's Canadian vice president, Geoffrey Hedges. Miele products have been available in Canada since 1989; they are more expensive to buy than North American machines, but their energy advantages make their life-cycle cost more and more attractive as energy prices continue to climb and government-imposed energy standards become more stringent. For example, most North American dishwashers have a life expectancy of five to seven years; Miele's dishwasher is rated for 10,000 hours of use, or about 20 years of normal treatment. According to Hedges, Miele dishwashers and clothes washers installed in Esprit, a 220-unit condominium project in Mississauga, Ontario, in 1991, resulted in an energy saving of 134,640 kWh per year—about $12,000 at current prices—most of that because they used 500,000 fewer gallons of water than standard North American appliances would have.

Both the refrigerator and dishwasher are powered by electricity, but the design team did not want to use an electric range—the energy consumption of most electric ranges is approximately 1,000 kWh/year, about the same as gas ranges when their consumption is converted to electrical units, but gas is cheaper (about $20 per year compared with $90 for electricity) and cleaner. They ran into a problem when considering ranges, however: the guidelines set by the Advanced Houses program specified that all gas-, oil- and wood-burning appliances had to be sealed-combustion units, which meant they had to draw their combustion air directly from the outside and also vent their exhaust gases directly outside. The problem was there were no sealed-combustion gas ranges on the market. The solution was to invent one.

They approached the Canadian Gas Research Institute (CGRI) for help. Researchers there took a conventional gas-range body and fitted it with a solid glass top of the kind found on some electric ranges. The glass top prevented combustion gases from escaping into the kitchen. They then sealed the combustion chamber under the glass top and fitted it with a small fan that exhausts the flue gases out of the house through a heat exchanger: the flue gases going out through the exhaust pipe heat up the colder air coming in, which improves the overall efficiency of the range. According to CGRI's Robert Lafontaine, this modified gas range is not available on the market yet, but it soon will be: "A lot of manufacturers have come up to take a look at it and have been suitably impressed," he says. "As soon as government regulations start taking a serious look at indoor-air-quality issues, manufacturers will have to start making sealed-combustion units. It's just a matter of time." Lafontaine says sealed-combustion units shouldn't cost much more than standard ranges, since the only real differences are the glass top and the fan.

In most houses, the biggest single electricity guzzler is the water heater. Ontario Hydro estimates that the average water heater uses 12,400 kWh of electricity a year, or about one-quarter of the entire electricity budget for the house. There are ways to lower that, even in conventional houses. Each heater has two electric coils in it, operated by individual thermostats that can be set between 120 and 170 degrees F. When the tank is installed, both coils are usually preset by the manufacturer at the highest setting, 170 degrees, probably for testing. Most people, however, are unaware that they can lower that setting themselves and simply go on adding cold water to their hot water in baths, showers, dishwater, and so on. This represents an incredible waste of energy. And although most new tanks have a few inches of fibreglass insulation under their metal shells, it's also a good idea to wrap the tank in an extra blanket of insulation: the amount of heat lost from a poorly insulated tank is about 900 kWh a year, the equivalent of having a 100-watt incandescent bulb burning 24 hours a day every day for a year.

That helps those stuck with electric water heaters. But the Greenhome designers didn't want electrically heated water. The Greenhome's water is heated by a combination of solar energy and natural gas. Two 4-by-8-foot solar hot-water panels on the south-facing roof (manufactured by Thermo Dynamics of Dartmouth, Nova Scotia) collect heat from the sun and transfer it to a glycol solution in the copper coils; the glycol runs down through insulated copper tubing to a 100-gallon hot-water storage tank in the mechanical room. The bottom half of the upright tank is heated by the solar panels; the top half is heated by a backup gas burner.

When the Thermo Dynamics preheat system was tested at the National Solar Test Facility in a simulated real-life scenario in which 80 gallons of hot water at 130 degrees F were required, the solar system was found to contribute up to 3,500 kWh of electricity savings per year. At that rate and with lowered hot-water requirements and the substitution of natural gas for electricity backup, the Greenhome's total hot-water-heating budget is just under 1,000 kWh a year, less than one-twelfth that of a conventional house and about the same amount as a poorly insulated hot-water tank *loses* in a year.

That makes it almost economical to fill the clothes washer from the hot-water tank, but the Greenhome doesn't even do that. Most clothes washers haven't changed much since the modified butter churns they were at the turn of the century: Thor Electric marketed the first elec-

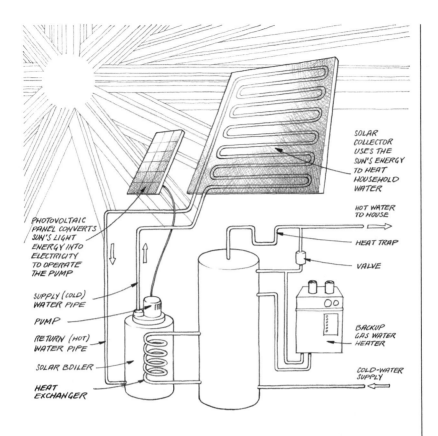

SOLAR COLLECTOR USES THE SUN'S ENERGY TO HEAT HOUSEHOLD WATER

HOT WATER TO HOUSE

PHOTOVOLTAIC PANEL CONVERTS SUN'S LIGHT ENERGY INTO ELECTRICITY TO OPERATE THE PUMP

HEAT TRAP

VALVE

SUPPLY (COLD) WATER PIPE

PUMP

RETURN (HOT) WATER PIPE

BACKUP GAS WATER HEATER

SOLAR BOILER

COLD-WATER SUPPLY

HEAT EXCHANGER

A solar collector on the Greenhome's roof heats a glycol solution in the tubes that is then pumped down to a heat exchanger in the mechanical room. There, the heat is extracted from the glycol and then transferred to the hot-water tank. As a consequence, 50 percent of the Green-home's hot water is actually heated for free.

tric clothes washer in the United States in 1909, and like the Canadian-made Acme B advertised in Eaton's 1922 catalogue ("Let Electricity do the hard work!"), it consisted of a wooden tub with a wringer and agitator hooked up to an electric motor. The wringer has disappeared, but not much else has changed: most modern automatic washers are still little more than agitators hooked up to electric motors, and they are still loaded from the top.

Top-loading clothes washers are simply not energy-efficient, and yet they dominate the North American market. Top-loading machines use more than twice the hot water that front-loaders use; heating that water electrically consumes far more energy than running the machine itself. An 8-pound load of laundry on regular cycle uses, on average, 38 gallons of water in a top-loading model; with front-loading machines, water use drops to 16 gallons. According to a recent article in *Consumer Reports*, an average family washes 182 loads of laundry a year; at that rate, the average top-loader would use 6,900 gallons of water a year, while the average front-loader would use only 2,900 gallons.

The most efficient front-loader sold in Canada is a Miele, the model chosen by the Greenhome design team. It uses only 14 gallons of water per 11-pound wash, or 2,500 gallons per year, and has other energy-saving features as well. A built-in water heater, for example, means it doesn't have to be hooked up to the hot-water tank; it uses only 312 kWh of electricity a year, compared with more than 1,200 kWh for most top-loaders, because its DC motor runs on less electricity than the AC motors in conventional washers. Less water also means less detergent. It goes through five rinse cycles, compared with 1½ for most machines, and its 1,600 rpm spin cycle – driven by an energy-efficient DC motor – removes 85 percent of the water from clothes, twice as much as other machines do, so the clothes dry faster and at a much lower cost if a dryer is doing the drying.

According to Geoffrey Hedges, the new American energy-efficiency standards imposed on North American appliances will be impossible to meet with conventional top-loading machines: "Within five years," he predicts, "all North American manufacturers will have to switch to front-loaders. It's the only way they can go."

While we're in the laundry room – actually, in the Greenhome, the laundry room is also the kitchen, since all five major appliances are located there to save space as well as plumbing and wiring costs – let's look at the dryer. Waterloo Hydro estimates that the average electric clothes dryer consumes between 360 kWh and 1,680 kWh of electricity a year; that's a wide variance, and it's worth taking a look at it. The EnerGuide directory lists 154 different models of electric dryers, ranging in energy consumption from 420 kWh per year to 1,128 kWh per year, depending on drum capacity and assuming 34 operations per month. Waterloo Hydro's estimates are based on observations of actual use; the low figure represents households in which dryers are not used every day of the year, and the higher figure represents houses in which the dryer is used nearly twice a day all year round. Obviously, homeowners have a lot of control over the amount of electricity their dryers consume. During the summer, for example, clothes can be hung outside to dry, and the Greenhome has an umbrella-type clothesline in the backyard for that purpose. In general, though, dryers use a lot of electricity—it's an appliance that uses electricity to generate heat.

The Greenhome has an Inglis natural-gas dryer equipped with a sensor that automatically turns the dryer off when the clothes in it reach a predetermined degree of dryness. Many electric dryers have this feature, and it's a good one in terms of energy use. Normally, dryers work on timers: you set the timer for a certain length of time, and you go off to do something else while the clothes dry. Almost inevitably, the dryer will continue to run long after the clothes are dry, simply because there is no way to guess exactly how long that will take.

The Greenhome dryer has a further energy option: as with the range, the gas burner in the dryer is a sealed-combustion unit, with intake and exhaust air passing through an air-to-air heat exchanger; outgoing air preheats incoming air, lessening the heating load for the appliance.

A good proportion of the Greenhome's energy budget has been calculated by using an expanded version of what is sometimes called life-cycle costing. Life-cycle costing is sort of the opposite of built-in obsolescence; instead of considering only the purchase price of an appliance, consider how much that appliance is going to cost to operate over its expected life span. Before the 1960s, appliances were expected to need repair or servicing at least once during their first year of operation and twice a year after that; during the 1960s, manufacturers realized that this repairing business was cutting down on sales of new appliances, so they began making their appliances irreparable—toasters riveted together instead of screwed, sealed plastic motors that had to be replaced. Consumer backlash against this built-in obsolescence has changed that policy, at least to the extent that most appliances can now be repaired. But the expected life spans of appliances have not been extended, at least not in North America. This means that we can calculate the life-cycle cost of two dishwashers, for example, without having to factor in repair costs.

On the face of it, the Miele dishwasher in the Greenhome is nearly twice as expensive as an average North American dishwasher. North American models run from about $500 to $1,100; Mieles start at $1,200 and go up to more than $2,400. But when we compare their operating costs over 10 years, the price tags more than even out. An $800 North American dishwasher with a life span of seven years will have to be replaced once in those 10 years; a Miele, with a life span of 20 years, will not. That makes the capital cost of dishwashing about even: $1,600 for two North American models compared with $1,500 for a midrange Miele. The differences in the operating costs are even more dramatic. On the basis of 10.5 gallons of water per wash, 950 kWh per year of electricity and an average increase of 8 percent per year in the cost of electricity, the North American dishwasher will cost $1,030 to operate over the next 10 years; the Miele will cost only $620. The environmental costs can also be factored in: at the end of those 10 years, there will be one more North American dishwasher in a scrap pile somewhere. As Lawrence Solomon notes in *The Conserver Solution*, "right now, the manufacturer benefits when he can keep his costs down. Once life-cycle pricing is adopted, the manufacturer will benefit when he can keep his customers' costs down. It's a small change in orientation, but its ramifications are sweeping." Solomon wrote that in 1978.

□ □ □ □

In the middle of December, before the insulation was in, the cold wind coming down from the north blew through the Greenhome's walls and around its windows, making work inside difficult. Everyone was bundled up; even the earflaps under Phil's hard hat were down. The five electricians who had volunteered to install the wiring were moving about stiffly, unrolling great lengths of white 12-gauge wire, drilling holes through the wall trusses and the interior stud walls, running wire from switch boxes and ceiling fixtures down to the main electrical panel in the basement mechanical room. Except for the grey wires that ran from every electrical outlet to the panel that would be hooked up to the Greenhome's monitoring system, there wasn't much difference between what the electricians were doing here and what electricians have been doing for decades. Still, they found a few things to grumble about.

To cut down on thermal bridging, the switch boxes in the exterior walls were plastic instead of metal and had neoprene flanges on them that could be sealed directly to the vapour barrier so that the outlets didn't have to be individually bagged. In my experience, electricians hate bagging electrical outlets, so this looked like a good system. But the electricians didn't like them. "You get used to working with one thing, and that's what you like to work with," said one of them, blowing on his bare hands. "Take those wall trusses, for example. They have 1-inch holes punched in them at the factory to run the wiring through, but look at them – the holes don't line up." I looked at them. The holes were all spaced a foot apart, from floor to ceiling, but it was true that they didn't line up; the wire running through the outside walls went up and down like lines on a seismograph recording a series of very, very slight earth tremors. I remarked that it didn't seem like much of an inconvenience.

"No," agreed the electrician, "it's not a big thing, but to us, it doesn't look good. If we'd drilled those holes ourselves, we'd've made them straight. Like I say, it's not a big thing, but it's the way we like to work."

Meanwhile, in the basement, a worker named Terry was installing the ductwork that would deliver warm air from the furnace to the house and cooled air from the house back to the furnace. The ducts came in the form of bent sheets of galvanized metal, which Terry had to fit together with prefabricated metal flanges and then attach to the basement ceiling. Between the rumblings of sheet-metal thunder that emanated from the hollow ducts as he worked, he explained how the Greenhome's duct system differed from that of traditional houses.

In most houses, including the Greenhome, the main hot-air duct from the furnace runs along the centre of the house, like an esophagus embedded in the basement ceiling under the floor joists: a long, rectangular galvanized box of sheet metal with numerous smaller, round sheet-metal ducts sprouting out of it like bronchial tubes. These bronchials deliver the warm air to the individual rooms of the house. In most cases, this main duct is the same size at the furnace end as it is at its farthest extremity. In terms of air flow, this is the second-least-efficient design possible (the least efficient is a funnel-shaped duct with the narrow end at the furnace). As air flows through a duct, its velocity decreases as its volume increases, and its pressure decreases with its velocity. In other words, the more slowly the air moves in it, the less pressure there is to force it up into the house. The result is that to keep a house warm, a big duct means a big fan. This principle was first articulated by the Swiss mathematician Daniel Bernoulli in 1738, but its significance to air flow and duct design has only recently been appreciated. The most efficient main duct is one that becomes progressively narrower the farther it gets from the furnace. The duct in the Greenhome is exactly that—a reverse funnel. Bernoulli would be delighted.

The shape of the main duct affects air pressure in the heating system, but so does the design of the smaller ducts that channel warm air from the main duct to the individual rooms. Whenever there is a 90-degree turn in one of these ducts—as there must be, to get the air from the horizontal to the vertical—air slams into the elbow, causing back-eddies that produce energy losses. John Kokko uses the analogy of a river making a sudden right-angled turn, losing speed, dropping silt and causing swirling eddies in midstream. Air does the same thing, and the pressure drop increases the strain on the fan. In the Greenhome's ducts, wherever there is a 90-degree turn, the duct is fitted with "turning vanes," metal fins that help the air make the turn without slamming into the end first.

In most houses with forced-air heating, the warm-air vents in a room

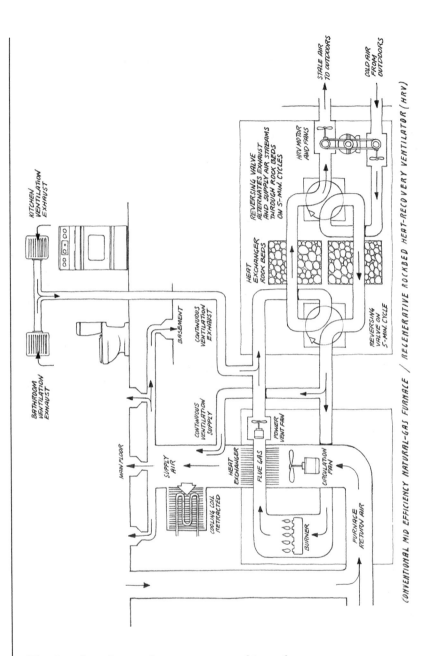

The Greenhome's space-heating system combines a heat-recovery ventilator (HRV) with a gas furnace: all air entering or leaving the house passes through the HRV's rock beds, which preheat air before it goes into the combustion chamber. The furnace achieves 90 percent efficiency.

154

are located directly under the windows, 6 inches from the exterior wall. The theory is that the windows are the weakest element in the house's insulation barrier; most of the cold air coming in via the wall comes in through or around the window, so that's where the heat is needed most. With a proper air barrier, good insulation and superwindows, this theory no longer applies, if it ever did: in reality, the theory requires such long ducts from the main one that by the time the air gets to the room, it isn't particularly hot anymore and, according to Bernoulli, doesn't have a lot of pressure left. The other, less obvious disadvantage is that the fan needed to force the air that far consumes a lot more energy than would be necessary with shorter ducts.

In the Greenhome, the warm-air vents are never located on an exterior wall; all of them are near interior walls, as close as possible to the main duct. This means the air is warmer when it enters a room and does not need a huge fan to get it there. The furnace fan required to push all this air around the Greenhome is extremely small—rated to move only 500 cubic feet of air per minute during the heating season, which Robert Lafontaine says is "surprisingly low," considering that most forced-air-furnace fans are rated at about four times that volume. The Greenhome fan is run by a new type of motor, manufactured by General Electric, called an electronically commutated motor, or ECM; it contains a small circuit that can be programmed to move different amounts of air at different times of the year. The Greenhome's furnace ECM will move 500 cubic feet of air per minute in winter and switch automatically to 1,000 cubic feet per minute in summer, when the cooler air requires more force to move it efficiently. ECMs are rated at 80 percent efficiency, compared with only 65 percent for conventional electric motors. What's more, adds John, "a conventional motor's efficiency decreases dramatically when you slow it down to 500 cubic feet a minute. Efficiency at that low rate could be less than 10 percent. With an ECM, that doesn't happen: its efficiency stays at 80 percent no matter what flow it's set at." In the Greenhome, ECM motors drive not only the furnace fan but also the cistern pump and the vent fan in the kitchen range.

In most houses with forced-air heating, the vents for delivering hot air to a room and the vents for drawing cold air out of the room are all on the floor. This doesn't make a lot of sense. In fact, it gives an extra and unwanted meaning to the term "forced air"; warm air rises, so when it enters a room at floor level, it rises to the ceiling until it is cooler than the new warm air, at which point it has to sink back down to the floor to be sucked out through the cold-air return vent. As it sinks, however, it mixes with the warm air that is on its way up to the ceiling and cools that air. In other words, with both warm-air and cold-air vents in the floor, air from the furnace is in the room twice as long as it needs to be and is never as warm as it ought to be.

On the upper level of the Greenhome, warm air enters the room through ducts in the floor and leaves through cold-air vents located at the top of the wall, right at the ceiling. As it enters the room, it rises to the ceiling and is drawn out before it can mix with the new warm air following it. On the lower level, the warm air enters at ceiling level, sinks as it cools and is drawn out through cold-air vents in the wall at the floor level. This design means that the warm air is in the room for only as long as it needs to be: it also means that it is still slightly warm when it is drawn back into the heat exchanger, where its heat helps to warm up the cold combustion air that is drawn into the furnace from outside.

Terry runs the two central duct systems—warm air in at the ceiling, cold air out at the floor—into the mechanical room and aims them at the far corner, where the garage wall meets the exterior wall. "That's where the HRV will go," he says, pointing to the corner.

"When will that be?" I ask.

"Just as soon as they invent it, I guess," he replies.

HRV—heat-recovery ventilator—is the current name for what was originally called an air-to-air heat exchanger; its name is really only a partial description of what is going to go into that corner of the mechanical room. HRVs were invented in Canada more than 15 years ago—the first one was built for Saskatchewan Conservation House in 1977—and they have since become one of the fastest-growing segments in Canada's low-energy industry: more than 50,000 units are sold internationally each year.

HRVs were the inevitable consequence of the evolution of energy-efficient houses; they're the answer to the question, Hey, now that we've got these airtight houses, how do we get fresh air into them and stale air out? So little air moves through the walls of a superinsulated airtight house—less than half an air change per hour—that fresh air has to be pumped in and stale air pumped out. Otherwise, cooking and, er, post-cooking odours, off-gassing from chemicals in paints, carpets, household cleansers, and so on, and moisture from showers, dishwashers and steam irons build up until the house becomes noticeably smelly, wet and—unhealthy. Some houses were so airtight that their owners couldn't keep a woodstove going in them; a dwindling wood fire in an airtight house is like a gasping canary in a mine shaft. No one wanted to simply open a window or stick a fan in a hole in the wall; that seemed to undermine the whole philosophy behind airtight construction. The idea was to stop uncontrolled air from entering a house through a thousand tiny ruptures in the envelope; concentrating incoming air through one big hole didn't seem like much of an improvement.

Enter the HRV. An HRV is that one big hole, but it is not uncontrolled. Warm, moist, smelly air goes out of a house through the HRV, and cold, dry, fresh air enters a house through the HRV; while the two kinds of air are inside the HRV, they exchange some of their properties. As its name implies, a heat-recovery ventilator takes heat from the warm air that is being exhausted from inside the house and uses it to preheat the fresh air coming into the house. HRVs have been viewed with some skepticism in the building trade and some trepidation among homeowners and owner-builders, but let there be no mistake: an airtight house has to have some kind of mechanical ventilation system, or it won't work, and an HRV is the best kind it can have.

There are two basic styles of HRV – cylindrical counterflow (or tube-in-tube) exchangers and square crossflow exchangers – but both work on the same basic principle. An HRV is a box, and inside the box are layers of thin polypropylene membranes, 22 to 150 of them, with half-inch spaces between each membrane. Warm, moist, odoriferous stale air from inside the house goes out through alternate spaces between the membranes, and cold fresh air from outside the house comes in through the box in the opposite direction through the other spaces. By the time both kinds of air leave the exchanger, the warm air is colder and the cold air is warmer. Even with the early HRVs, recovering 65 to 90 percent of the heat from outgoing air and transferring it to incoming air, the potential fuel savings were considerable. When that warmed-up incoming air was connected to the intake on furnaces, furnaces no longer had to heat air from below zero to 70 degrees F in the winter; by the time outside air entered the furnace, it was already nearly 50 degrees. It takes a lot less fuel to heat air 20 degrees than it does to heat it 70 degrees.

There were problems with stand-alone HRV systems, however. For one thing, they were expensive to install because they required their own duct systems and fans; installing an HRV system in 1987 increased the cost of a house by up to $5,000, a cost that at current fuel prices, could have taken up to 30 years to recover. And then there was the problem of frosting. Warm air entering the HRV from exhaust vents in the kitchen and bathrooms was not only warm, it was moist, and as it cooled, the moisture would precipitate onto the membranes in the HRV. When temperatures dropped during the deepest, darkest days of winter (when HRVs were most used and most needed), the frigid incoming air would freeze the moisture on the membranes, reducing their efficiency and

157

sometimes clogging them up altogether. This frosting problem was solved by fitting the HRV with an electric defroster to preheat incoming air before it went into the HRV to be preheated again, not a very good solution from an energy standpoint.

At the same time that HRVs were frosting up and homeowners who had done an insulation retrofit on their houses were scratching their heads trying to find room for yet another set of ductwork (heating, cooling and now ventilating), inventors were beginning to toy with the novel idea of combining all three mechanical systems into one integrated whole. A century ago, a house had one ventilation system – leaky walls – and one heat source – a woodstove. Now, there wasn't only a central heating system, there was a central water heater, a central cooling system and a central ventilation system, each with its own fuel source, its own ducts and its own fans, sitting in its own corner of the basement. Clearly, all those systems couldn't be "central" as long as they were all separate. Why not combine them all into one unit so that the same fire that heated the house also heated the water; the excess heat given off by the water heater could be blown into the house; the same fan that pushed hot air into the house in winter could push cool air into it in summer; and the ducts that fed the furnace also filtered air through the HRV?

Such an integrated mechanical system (IMS) might be possible if everything were driven by electricity, but the energy savings were hardly worth the effort. Combining an HRV with a conventional combustion furnace wouldn't work because the furnace exhaust would go up the chimney and the ventilator exhaust would go out through the wall. The way to make an IMS most efficient would be to direct exhaust gases from the furnace through the HRV, thus recovering some of their heat, which would also help to solve the frosting problem. Exhaust gases from conventional furnaces, however, were far too hot: even in a relatively efficient gas furnace, flue temperatures were normally around 500 degrees F. For mechanical systems to become completely integrated, flue temperatures had to be lowered considerably. And that wasn't possible until mid- and high-efficiency gas furnaces came along.

Ordinary gas furnaces operate at a fuel efficiency of about 56 percent, mostly because so much heat escapes up the chimney but partly because they have pilot lights that consume extra fuel and because the dampers in the flue remain open all the time; even when the furnace is off, warm air is sucked up into the chimney, causing the furnace to come on more often. Moderate-efficiency gas furnaces address the smaller problems by using electronic ignitions and electrically operated damper controls that increase efficiency rates to about 65 percent. Flue temperatures, however, are still too high to be directed out through a conventional HRV.

Then came furnaces with fans that force-fed air into the combustion chamber as well as heat exchangers in the plenum that reduced flue temperatures and increased the furnace's efficiency to 75 to 80 percent. Moderate-efficiency furnaces cut heating bills up to 30 percent in most houses, but escaping flue gases were still so hot they had to be exhausted through a chimney; if directed through an HRV, they would heat up the stale airstream so quickly that its moisture precipitated in the HRV, resulting in a bigger frosting problem than ever.

High-efficiency gas furnaces incorporated several heat exchangers within the furnace itself so that flue gases were less than 100 degrees F, so cool that a chimney was unnecessary: exhaust gases could be vented directly out through the side of the house like those from a clothes dryer. And anything that cool could be directed through an HRV first. Suddenly, IMS began to make sense.

High-efficiency gas furnaces are so efficient, in fact, that they don't really need to be integrated with another source of heat—like an HRV—to be made more efficient. The trick was to find a way to increase the efficiency of a midefficiency gas furnace, and when John Overall at CGRI started tinkering with the problem in 1988, that's what he was trying to do: combine a midefficiency gas furnace, a cooling system and an HRV in a way that did not require expensive condensing equipment but would still achieve high levels of energy efficiency. He developed "a gas-fired integrated appliance for heating and ventilating" that combined combustion products from a midefficiency gas furnace with the airstream vented from the building so that all air entering the HRV was diluted by a factor of 10 to 1, which means that the temperature drop was so gradual that it did not cause excessive condensation.

He also experimented with a new type of HRV; in fact, it was based on an even older concept borrowed from active-solar days. "Instead of exotic materials for heat recovery," he said at an early Greenhome meeting when he was introducing his prototype unit, "we recover heat in two regenerative beds filled with high-tech gravel." High-tech gravel?

"Well, beach pebbles, actually," says Robert Lafontaine, who brought the unit to the Greenhome in December. "Or maybe crushed limestone, I'm not sure. Anyway, it's quarter-inch aggregate."

"We looked at various types of plastic and metals for the beds," Overall said at the meeting in March, "but when we cranked out all the calculations, the answer came out gravel. I don't mind saying it irritated the hell out of us, but that's what the figures showed, so we went with it. It seems to work."

159

The principle is quite simple. As indoor air leaves the house, it passes through a small, screened metal box full of stones; on its way through, it warms up the stones. Outdoor air is then drawn in through a valve system inside the unit that directs the air into the stone bed to be warmed up before it flows into the combustion chamber. The valve system is necessary because there are actually two stone beds; while one bed is being cooled by incoming cold air, the other is being warmed by outgoing warm air. Just before the cooling bed reaches the freezing point, the valve switches the air flow so that cold air comes in through the warmed box and indoor air leaves through the cooled box. That way, there is no chance of a freeze-up in either box.

The HRV recovers about 83 percent of the heat from the air that enters it; the air comes in at 480 degrees F and leaves the recovery beds at about room temperature. The heat extracted by the HRV boosts the overall efficiency of the CGRI appliance to well over 90 percent, which is at least as efficient as the much more expensive high-efficiency furnace. The total amount of natural gas required to heat the Greenhome will be less than 500 cubic metres a year, the equivalent of about 1,000 kWh of electricity. This is less than 6 percent of what Ontario Hydro estimates most southern Ontario houses spend on space heating: 1,000 kWh instead of 18,500. Not bad.

At the design meeting, John Overall passed around some photographs of the prototype furnace – they showed a sheet-metal box about the size of a dishwasher with a few round ducts running into or out of it, it was hard to say which. "Can I come down to CGRI and have a closer look at it?" John Kokko asked.

John Overall cleared his throat. "We thought we had a deal with a Canadian manufacturer to put this thing into production, but that seems to have fallen through," he said. "I really wouldn't like the details of how it works spread across the country until we line up another manufacturer and are ready to put the unit on the market."

John Kokko looked as though he was used to being suspected of industrial espionage. "What if I take an oath of secrecy or something?"

"We could show you the box."

"What about inside the box?"

John Overall cleared his throat again. Then he continued, "If we don't find a new manufacturer in two weeks, we'll start work on it ourselves," he said. "At that time, we can try to accommodate your questions and specifications."

When Robert Lafontaine showed up at the Greenhome in late December with the unit in the back of his CGRI van, I asked him about the manufacturer as I helped him carry the 80-pound regenerative beds into the basement.

"CGRI has signed a licensing agreement with a firm called DMO Industries, in Wallaceburg, Ontario," he said. "They're building a few more prototypes and doing some field evaluations before getting into it full tilt. Sort of testing the waters. But it looks good."

"What are you calling it?"

"We call it an integrated heating-and-ventilating appliance," he said.

"An IHVA," I said. "Has a nice ring to it."

"Yes," said Robert. "We thought so."

On the 10th of February, John and I were sweeping up in the Greenhome. The drywall hadn't been put up yet, and some of the cellulose insulation had trickled down from the walls onto the floor through the nylon fishnet. We were sweeping it up so that when the vapour-barrier installers came in, they could caulk the poly on the walls to the flap of poly that poked out from under the bottom plate. The temperature outside was minus 5 degrees F, but the house, heated only by the gas fireplace in the great room, was perfectly comfortable.

The sealed-combustion fireplace is fitted with a thermostat that turns it on when the ambient temperature in the house falls below 59 degrees F and off when the room reaches 68 degrees, because, as John explains, "the thermal output of the fireplace is greater than the heat load of the house." In other words, the whole house could be heated by the fireplace, but that is a much less energy-efficient way of space heating because the flue gases are not integrated with the HRV/furnace in the basement. The fireplace chimney has an electronically operated damper in it that closes the flue as soon as the fireplace shuts off, to prevent warm air in the house from being sucked out through the chimney.

Werner came in to see how we were doing. Now that the main construction phase was over, he had more or less handed the site supervisor's role over to John. He seemed visibly relieved; he had steered the Greenhome through five months of complex negotiations with volunteer and paid workers, suppliers, installers, consultants and hangers-around like me, and he seemed glad to be finished. All that was left to do now was to oversee the installation of the air/vapour barrier and drywall, to hook up the appliances and to decide on the kind of paint, the carpets and curtains, the trim and the plumbing fixtures.

When we finished sweeping, we leaned the brooms in a corner of the great room and looked around. The 2-by-4 framework around the fireplace and chimney hadn't been drywalled yet. Instead of being trucked off to the dump, the off-cuts of drywall from the ceiling were stacked between the framing and the metal fireplace; they would remain there to act as thermal mass when the frame was closed in. The grey cellulose walls looked warm and comfortable, as if we were standing inside the belly of a huge stuffed whale. Although it was bitterly cold outside, we were warm. There wasn't a trace of frost or condensation on the windows. The fireplace hissed away quietly and efficiently. The house was working, but not as a machine works; rather, as a living thing works.

FINISHINGS

□

THE STRENGTH OF A NATION IS DERIVED FROM THE
INTEGRITY OF ITS HOMES.
    —CONFUCIUS

The roof coverings settlers put on their shanties and log houses 200 years ago were very sensible adaptations of the clay tiles common in southern Europe. Called troughs, they were made from whole logs split down the middle, hollowed out and laid vertically up the pitch of the roof. Catharine Parr Traill describes a shanty roof in *The Backwoods of Canada* (1836): "The roof is frequently composed of logs split and hollowed with the axe and placed side by side so that the edges rest on each other; the concave and convex surfaces being alternately uppermost, every other log forms a channel to carry off the rain and melting snow. The eaves of this building resemble the scalloped edges of clam shells, but keep the interior dry, far more so than roofs formed of bark or boards."

Early settlers also used shakes and shingles on their roofs, not only out west, where the western red cedar (*Thuja plicata*) still provides the best shingle wood, but also in the east, where pine, white oak, red oak and the eastern version of cedar—actually a species of juniper called *bâton rouge* by the French because it looked like cedar—were acceptable substitutes, except that pine contains pitch and thus has a greater tendency to burst into flames, a decided disadvantage on a house heated by a woodstove or fireplace.

"Shingles," by the way, defined as thin pieces of wood having parallel sides and one end thicker than the other, seem always to have been made from wood. An English text published in 1308 describes a roofing method in which "the lathe . . . is nailed thwarteouer to [across] the rafters, and thereon hongeth sclattes, tile and schingles." Shingles and shakes are made today pretty much the way they were in 1308; shingles are cut with a saw and are more uniform-looking than shakes, which are split or hand-riven radially from the log. Shakes are rougher, more rustic than shingles, which to some people is a plus, and because shakes have no sawn edges, they also shed water better, which means they last longer.

Wood shingles are expensive, however, and labour-intensive, and although their life-cycle cost is relatively low, there just aren't enough trees left to put roofs on the hundreds of thousands of houses built in North America each year. By far the most common roofing material today is the asphalt shingle (which we pronounce, for some arcane reason, "ashfault"). And it will probably come as no surprise to you by now that from

164

an energy-use standpoint, asphalt shingles are the worst choice you can make. What you may not know, however, is that asphalt shingles don't make much sense from an economic standpoint either. Although they are one of the least expensive roofing materials available—current costs are about $2.50 per square foot of roof, not installed—they also have the shortest life span—about 15 to 20 years, depending on the thickness of the shingle and whether it has a cardboard or fibreglass base. This means that over the 60 years that life-cycle costing takes into account, an asphalt roof will have to be replaced or recovered at least three times, probably four, which raises its long-term cost to more than twice that of any other roofing material.

But aside from the out-of-pocket cost to builders and homeowners, asphalt shingles have a high environmental cost. They are made, after all, of the same stuff we put on our roads—a bituminous by-product that requires an enormous amount of energy to produce, has no recycling potential whatsoever and does not biodegrade when hauled to a land-fill site, as an extraordinary amount of it is every year. The Greater Toronto Home Builders' Association estimates that asphalt shingles make up 6 percent of all waste from new-house construction. On the basis of 140,000 housing starts in Canada in 1992, that makes 25,000 tons of asphalt dumped into Canadian landfill sites from new houses alone. Add to that the waste from all the 15-year-old houses that are being reroofed every year and then the waste from all the 30-year-old houses, and the total amount of nonbiodegradable petroleum-based bituminous asphalt going into our dumps becomes truly staggering: according to the Canadian Housing Information Centre, Canadian manufacturers produce about 33 million bundles of asphalt shingles a year, each of which weighs 75 pounds and costs about $10; if 6 percent of those are wasted and dumped, that means Canadians spend more than $20 million a year carting asphalt to landfill sites.

When the Greenhome design team discussed roofing materials, it didn't take them long to eliminate asphalt shingles. They discussed cedar shingles but decided that because they come from large, first-growth trees and would have to be transported all the way from British Columbia, they would involve too much embodied energy. But that still left a lot of other possibilities. John drew up a chart of all the alternative roofing materials, factoring in the life-cycle and environmental costs of installation, maintenance and eventual replacement, calculated those costs over a period of 60 years and brought the list to the meeting.

From the other side of Westvale Drive, the Greenhome's roof appears to be covered with dark clay tiles whose graceful, undulating curves are reminiscent of sun-baked Mediterranean villas. When architect Richard Stein, author of *Architecture and Energy*, looked out of an upper window in Siena, Italy, his eye was soothed by the "large and varied landscape of rooftops, all in terra cotta tiles, all slightly different in colour through different firings and lichen growths." A great deal of that pleasure came from the knowledge that he was looking at what others have called "appropriate architecture," building methods and materials that seem to spring out of the very ground on which the buildings sit. The clay for the tiles and bricks and the lumber for the rafters and trim came from within carting distance of the houses they ended up in. The shape and style of the houses, in fact, were determined by the local building materials available to their owners. Eighteenth-century Canada was rich in trees, not terra cotta, and so our houses were built with logs and shingled with wood.

The Greenhome roof has the soothing effect of clay tiles, but it's made from material a lot closer to hand in Waterloo than clay. In fact, the Greenhome roof is made of 26-gauge zinc-coated galvanized barrier steel.

Metal roofing has been around for a long time. Most of us associate it with the old corrugated material we still see on barns and country sheds, rusted and bent, flapping in the wind; those roofs have been there for a long time, and look it. The new metal roofs don't rust, however, and they'll usually last longer than the building they're put on. Their life-cycle cost is what makes them attractive. "Having been raised in a society where the change from metal roofing to asphalt shingles was considered a sign of progress," writes Gene Logsdon in *The Low Maintenance House*, "I am surprised when every roofer I ask says that metal roofs are the best buy for the money of *any* kind of roofing."

The steel roof on the Greenhome is called Tile Master. It is Stelco G90 galvanized steel, covered on both sides with zinc and on top with an 8-mil film of plasticized vinyl coating to protect the metal from corrosion; it was developed as a siding for industrial buildings, and Stelco's brochure says it "has been used extensively in chemically aggressive environments such as those found in pulp-and-paper mills, potash refineries, smelters and chemical plants, demonstrating its superiority in areas where the elements in combination with salts, acids, alkalies, oxidizing agents and industrial and automotive fumes are quickly deteriorating a variety of other building materials."

"This is tough stuff," Doug Summerhayes told John and Werner at the Enermodal offices in September. "It will endure for a lifetime." In fact, Mair Roofing, the installation company Summerhayes works for, guarantees that it will: as long as the roof is inspected every two years, any replacement or maintenance necessary is free to the owner. "We've back-tested it outside for 15 years and found no change, and our lab checked it for colour deterioration and calculates that over 50 years, it will have a colour loss of less than 5 percent." Apart from soothing lines and tough-ness, however, Tile Master steel roofing has other advantages. Each 15-by-56-inch panel, pressed to look like a row of tiles, is attached to the roof decking over 15-pound felt paper with only six screws, which are then caulked with silicone sealant. "We prefer to screw it to plywood," said Summerhayes, "but aspenite will hold almost as well." The smooth vinyl surface will not build up mould or mildew as asphalt and cedar shingles tend to do. And the airspace between the tile and the decking, absent under asphalt shingles, keeps the decking and felt paper 16 Celsius degrees cooler than they would be under asphalt. That has enor-mous implications for summer air conditioning. "And," Summerhayes added, "it's recyclable. Any off-cuts or ends returned to us go back to Stelco to be melted down and recast. And even when a house is torn down, we can salvage the shingles and use them somewhere else."

□ □

WHEN WE BUILD, LET US THINK THAT WE BUILD FOREVER.
— JOHN RUSKIN

One of the strongest contenders for the Greenhome's roofing mate-rial was a product called Rustic Roofing Shingles, manufactured in Que-bec by Canadian Pacific Forest Products (CPFP). These shingles look like cedar shakes and even act like cedar shakes: each shingle is 12 inches by 48 inches, and although they are the colour of cedar when installed, they weather to what the brochure calls "a beautiful silver-grey" over the first few years. They come with a 25-year warranty, which is longer than that for asphalt or fibreglass shingles, and they have one great advan-tage over cedar shingles or shakes: they are made of recycled wood.

The product was hard to pass up—it was innovative, it was Canadian, it was recycled—and so the design team decided to use it as siding in-stead, on the Greenhome's gable ends and under the windows. After all, cedar shakes are used as siding material too. The team also chose another Canadian Pacific Forest product called Classic Smooth Prefin-ished Lap Siding, from CPFP's Canexel Division, for the rest of the ex-terior walls. The siding is made of the same fibrous recycled material as the shingles. At a meeting in November, when the construction pro-cess was in full flight, Ian Cook brought a sample of it to the Enermodal offices to show Werner, John and Steve. The panel was passed around the table: it had a sort of greyish green finish baked on the outside, but the underside was brown and shiny and smooth, and the fibrous material in between looked exactly like:

"Masonite," Steve said wonderingly. "This stuff looks like Masonite."

"Yes, well," said Ian, "it's called hardboard now." Hardboard is essentially a finer-fibred aspenite. It is made from lignocellulosic fibres, "usually wood," says the brochure, "not only from logs but also to an ever-increasing degree from wood shavings and leftovers from other operations" – in other words, sweepings from the forest floor. These are ground up into ⅝-inch chips and mixed with stiffening agents, then forced together under pressure in a gas-fired kiln to a density of at least 600 kilograms per cubic metre. Hardboard itself, as Steve's comparison underlined, is not new; Canadians use a billion square feet of it every year, in thousands of different products, from prefinished wall panels to the bottoms of dresser drawers to the matrix under integrated computer circuitry to the stuff under the nice wood veneers in certain inexpensive brands of Scandinavian-style furniture. Using it as a base for shingles and siding is, however, a recent development. The main advantage for builders (as distinct from environmentalists) is that its density gives it a strength and a hardness that outperform all other kinds of siding except brick. Its "modulus of rupture," or MOR – an engineering term for the force of the blow a karate expert would need to chop a piece of it in half – is more than twice the industry standard (4,180 psi); its hardness, that is, its ability to withstand denting and scratching, is even more impressive. This is measured by a rather quirky procedure called the Janka ball test, in which a steel ball 0.444 of an inch in diameter is placed on the surface of the board, and the amount of force required to drive the ball into the board to a depth of half its diameter is measured. The Canadian Government Standard for this test is 600 pounds; Canexel sidings average 2,900 pounds. A ball driven with that much force into aluminum siding would probably end up in the living room.

Aluminum siding has, shall we say, other disadvantages. The embodied-energy question is a bit muddied in the case of aluminum and is worth looking at afresh. As was mentioned in an earlier chapter, more than 60 percent of all the aluminum beverage cans manufactured in North America are recycled, and recycled aluminum requires only 5 percent of the energy it takes to make new aluminum. So far, so good. The problem is that all the recycled aluminum beverage cans are used to make more aluminum beverage cans; none of that recycled material is used to make any other kind of product, including aluminum siding.

If you ask a recycling company why that is, you will probably be told that the aluminum in beverage cans is the wrong alloy for making siding. In fact, sheet-aluminum manufacturers would love to get their hands on aluminum beverage cans. "Pop cans would be perfectly okay for siding," says Walter Kellerman, a research scientist at Alcan's lab in Kingston, Ontario. Then he goes on to add, "It's just that they're worth more to the industry as pop cans. It's a money game." Recyclers, according to Kellerman, can sell pop cans to the pop-can industry for 44 cents a pound, but they'd get only 34 cents a pound if they sold them to sheet-aluminum manufacturers.

Aluminum siding is made from 100 percent raw aluminum. Kellerman says that all of Alcan's in-house aluminum waste from the manufacture of window frames, door trim, and so on, is melted down again, but virtually no recycled aluminum from the building industry goes back to the manufacturers to be made into siding — not even old siding or roofing material removed from demolished buildings, of which there are tons and tons available annually. Most of that is recycled, says Kellerman, but not as aluminum siding. "The wrecker sells it as 'old sheet' to secondary scrap-metal dealers, who lump it in with other kinds of metal, such as 'mixed low copper' and 'auto-shred,' which, as its name implies, is what's left of a car when all the iron is removed, and this mixture is sold in bales to factories that make foundry castings." Foundry castings are the moulds that are used to make other metal products, such as automotive cylinder heads. "Anything can be used to make foundry alloy," says Kellerman. Foundry-casting companies pay the scrap dealers 34 cents a pound for foundry alloy; if they separated it out at their yards, they could get more for it as a distinct metal — 44 cents for aluminum, 39 cents for low copper, et cetera — but it would cost them more to separate it than they could get for it, so they don't bother. "From a research point of view," says Kellerman, "we would love to find a way to make new siding out of old siding. But there are hundreds and thousands of secondary scrap dealers out there, and trying to change their way of handling old sheet is impossible. The best we can do is to look for a way to separate it after we get it as foundry alloy, but so far, we have not been able to do that economically."

The main competition for aluminum siding in North America is vinyl siding, and from an environmental point of view, it is not much different from aluminum. It is made from thermostatic plastic and, like all plastic, is a petroleum by-product; vinyl can be recycled, and as we shall see, there is some recycled plastic in the Greenhome, but as is the case with aluminum, new vinyl siding is not made from recycled vinyl, and by and large, the design team tried to avoid plastic wherever possible. They didn't even consider vinyl siding.

◻ ◻ ◻

I RESPECT WASTE. WASTE MADE AMERICA WHAT IT IS
TODAY.
—RUSSELL BAKER, AMERICAN NOVELIST AND ESSAYIST

Reducing construction waste is one thing; after the house is built, however, it is equally important to reduce household waste. House construction in Canada in 1993 will probably send about 350,000 tons of waste to Canadian landfill sites, and according to recent statistics, Canada and the United States are also adept at producing waste *after* the house is built: fourth and fifth in the world, respectively (3.7 pounds per person per day in Canada, 3.3 in the United States; the leader is Australia, at 4.2, then New Zealand and France are tied at 4.0). This means the average North American household produces about three-quarters of a ton of garbage per person per year, which in turn means 18 million tons of solid waste goes to Canadian landfill sites and incinerators each year, unless it is recycled first.

Fortunately, a lot of it is. One of the specifications imposed by EMR on all Advanced Houses was that they contain recycling centres, special places set aside for sorting and storing recyclable household waste. The waste would be of two types: kitchen waste that would eventually be taken outside and placed in the compost bin (also an EMR specification), and the plastic, glass and paper waste that would go into the Blue Box for curbside pickup.

The Blue Box recycling program, now ubiquitous throughout North America (in 1991, there were 3,955 curbside recycling programs in the United States, serving 71 million people), actually began in Kitchener, Ontario, in 1980, when a company called Resource Integration Systems Ltd. (RIS), in tandem with Laidlaw Environmental Services Ltd., Canada's biggest waste-management company, launched an experiment in recycling in 1,200 Kitchener households. RIS divided the total number of households into four quadrants and used a different carrot in each quadrant to see which one would attract the most recycled material. They used promotional literature in the first zone; different pickup days in the second; community-action committees in the third; and in the fourth, something RIS president Derek Stephenson had seen work in Kelowna, B.C., a few years before: a blue plastic box that residents could haul out to the end of their driveways with the regular garbage. RIS workers stencilled "WE RECYCLE" on 250 boxes. When the experiment was over, the blue-box quadrant had out-recycled the others by a factor of three to one. The Blue Box was born. "That message on the boxes was almost as insightful as the box itself," Stephenson says. "People want to make a declaration."

In the United States, where volunteer recycling programs outnumber government-legislated ones by 1,569 to 1,376, 14 percent of all household solid waste is now recycled. The state with the best recycling record is Washington, where 34 percent of household waste is recycled. Although curbside recycling has had a huge profile in Canada as well, the figures are less impressive. In Ontario, which has the worst waste problem in Canada, only one out of every four municipalities (totalling 66 percent of the population) has a Blue Box program, and Ontario Multi-Material Recycling Inc. (OMMRI), a manufacturers' organization set up to manage the province's $60 million recycling program, estimated in 1991 that only about 7 percent of the province's residential solid waste is diverted from landfill sites: this figure included about 1.3 billion of the 2.1 billion pop cans sold in Ontario each year (pop cans represent only 1 percent by weight of Ontario's solid waste) and a little over 500,000 tons of other trash kept out of the province's landfill sites.

That isn't bad, but we could do a lot better. In fact, Ruth Grier, Ontario's energy minister, means to see that we do. In 1991, Grier announced her plan to have 90 percent of Ontario households on a Blue Box program by 1995 and to make recycling mandatory in all the province's offices, schools, restaurants and industries. Grier might be taking her cue from the state of Washington. There, the city of Seattle charges residents directly for curbside garbage pickup: $10.70 for each 20-gallon container, but recyclable items are picked up free of charge. As a result, Seattle's record is the highest of any city in North America: 45 percent of its garbage is recycled.

The Blue Box program has its critics, to be sure. No one doubts that we have to reduce, reuse and recycle, but some feel that Blue Boxes place almost all the emphasis on the latter at the expense of the first two: there is a tendency for us to feel we don't need to reduce the amount of goods we consume because we are recycling; we don't have to ask Coca-Cola to put its beverages in refillable containers, because we are recycling the cans; it's all right that liquor and wine producers don't have to spend millions of their own dollars to set up a deposit-and-return system for their bottles, because we are spending millions of our own dollars for curbside recycling programs (the $60 million that

OMMRI distributes comes in equal proportions from manufacturers, the province and the municipalities).

A second criticism isn't really a criticism of the recycling program but rather of what happens to the recycled material after it's removed from the Blue Boxes. A lot of it is reused, but an awful lot of other recycled material isn't. Almost all the green glass that has been collected since 1980, for example, is still sitting in stockpiles, waiting for someone to think of a use for it. Stacks and stacks of old newspapers are gathering dust and mould somewhere. Seattle, for instance, used to be able to sell its recycled newspapers for $10 a ton; last year, it had to pay $10 a ton to have them hauled away.

David McRobert, a policy analyst with the environmental lobby group Pollution Probe, has recently pointed out that changing Ontario's paper-products industry from a raw-resource-based economy to a secondary-materials economy (using recycled paper instead of trees to make new newsprint) will have a major impact on Ontario's northern communities: 41 towns in northern Ontario depend for survival on the forest industry, which represents 63,000 jobs and $10 billion a year. McRobert calculates that up to 25 percent of those jobs would be lost if half of Ontario's paper is made from recycled material. But McRobert also points out that there is no reason those jobs couldn't be replaced by the new "green industries" that will spring up in response to Ontario's recycling targets: recycled paper could be shipped north to be turned into new paper, and recycled wood products could also be hauled to northern Ontario chipping mills to be made into fibreboard. Forest-products companies could be encouraged to invest more of their profits in research on and development of new secondary-materials products. Canada currently has one of the worst forestry research-and-development rates in the world—about 0.5 percent of forest-products income is reinvested in research and development in Canada, compared with nearly 20 percent in Sweden. A happy exception to that rule is MacMillan Bloedel, which spent $150 million developing its Parallam engineered-wood division. If more builders follow Greenhome's example in finding innovative uses for engineered wood, then perhaps more forest-product companies will follow MacBlo's lead in providing it.

Research and development in the plastics industry has also turned up some surprising products. One of them is made of the controversial plastic nonrefillable beverage bottles, the kind that can be placed in Blue Boxes. The bottles are made from a tough, durable plastic called polyethylene terephthalate (PET), which is so strong it has all but replaced nylon in the manufacture of car tires and automotive seat belts. It's a good thing the bottles are recyclable, because North Americans discard 2.5 million soft-drink containers *every hour*. PET containers make up 6.5 percent of all landfill waste by weight, and more than that by volume. The controversy derives from the fact that environmentalists would much rather see beverages in refillable glass bottles, while soft-drink manufacturers are happy to let someone else pick up their containers and figure out what to use them for. "You can't sell what the public does not want," Stuart Hartley, executive director of the Ontario Soft Drink Association, said in a 1991 *Harrowsmith* article about the Blue Box dilemma. "Convenience is in our psyche, and again and again, our studies show that the weight of the container is a primary consideration." According to the soft-drink people, no one wants to lug home a heavy glass bottle when they can get a light PET container or an aluminum can — an odd conclusion to draw, since those heavy glass bottles haven't put much of a crimp in the wine and liquor trade, and sales of beer in refillable bottles far outstrip sales of beer in cans.

Still, 215 million pounds of PET are recycled every year — about 25 percent of it from bottles collected in Ontario and Quebec. Most of it is used to make new beverage containers, paintbrush bristles, package strapping and fibrefill for sleeping bags and parkas. And every year, Image Carpets Inc., a company based in Armuchee, Georgia, takes 50 million pounds of it and turns it into carpets. They call it the Enviro-Tech Carpet System; the bottles are cut into ¼-inch chips, and the chips are melted down and extruded through moulds that create fibres about the thickness of a human hair; because bottle-grade PET has to have low moisture absorption, carpet fibres made from it are stronger and more stain-resistant than any other kind of polyester resin, including nylon. For example, PET fibres have less than 0.5 percent moisture regain, whereas nylon will regain up to 6 percent. Image's Wearlon HT (for high twist) carpeting installed on the Greenhome's basement floor was tested by Image and found to successfully repel staining by ketchup, Kool-Aid, red wine, several types of cola, grape juice, mustard and blackberry jam — all without chemicals to make the carpet stain-resistant.

The full line of Image's Enviro-Tech carpets is distributed in Canada by Venture Carpets in Toronto. Jeff Sellner, of Sellner Contracting, the company that installed the Wearlon carpet in the Greenhome, says the carpet "looks fantastic and feels like any other carpet you'd find anywhere — saxonies, loops, you name it." Sellner is equally enthusiastic about the underlay that went beneath the carpet; it's made by Dura Undercushion from recycled car tires, and "it's the heaviest underlay I've ever seen in my life," he says.

Several recycling companies are making other kinds of products. The Greenhome used one called Superwood, between the basement concrete floor slab and the foundation wall to form a thermal break that will not rot away, ever. And John visited a demonstration project in Boston recently that was a house made almost entirely out of recycled plastic: even the shingles on the roof were made of old computer housings, he said with a kind of wonder in his voice.

Recycling is beginning to make sense to a lot of manufacturers, as John Langdon of Laidlaw Environmental is aware. Langdon's job is to find markets for the waste products that Laidlaw collects. If he can't find them, sometimes he creates them. Three years ago, he noticed that "post-consumer paint" was forming a significant proportion of the waste stream at Laidlaw's municipal hazardous-waste depots. Post-consumer paint is a euphemism for those drip-sided pails of half-hardened paint we all have in our basements, with the lids sort of pressed on loosely, which we are not allowed to put out with the trash or take to the dump. Many municipalities now have permanent hazardous-waste depots for such things — paint, old batteries, cleaning solvents and such — and some people actually bring such things to them. Langdon decided to look for a better way to dispose of the paint than the two traditional methods: incinerating it or "pugging it up" — mixing it with fly ash to turn it into a solid and then burying it in a secure landfill. "Paint is a valuable commodity," he says, "and it seems a shame to waste it; besides, anything we can keep out of the landfill is a plus." He initiated a pilot project with the Canadian Paint and Coatings Association to see whether it was technically possible to turn old paint back into new paint, and when they found out that it was, Laidlaw built a paint-recycling facility in Mississauga that began salvaging old paint by literally squeezing the buckets dry and filtering the paint that ran out of them. The next step was to melt down the buckets into steel billets. The billets were then sold to scrap-metal dealers, while the filtered paint was put into drums. Meanwhile, Langdon put out a call to paint companies to come and haul the recycled material away.

Surprisingly, the only company to answer the call was Scarfe Paints Ltd., a small family-run business in Owen Sound, Ontario. Scarfe's paint chemist, Ron Coates ("He's really an alchemist," says Langdon) took the recycled raw material and created four types of paint with it: a 100 percent recycled interior latex, available in off-white only; a 90 percent recycled interior latex that comes in 30 colours; an exterior latex that is 50 percent acrylic and 50 percent recycled, available in 60 colours; and a "block filler" — a base paint that smooths over concrete blocks to prepare them for a finish coat—also 100 percent recycled. "We've only been doing this for a year," says Coates, "but already our biggest problem has been supply." Laidlaw's facility recovered only 27,000 gallons of raw material in its first year of operation, but as Langdon says, "It's all just sitting there, ready to sell, so as soon as Ron gets enough orders, we'll be happy to ship it to him."

Scarfe already has several orders. Its first contract was with The Body Shop, an environment-conscious cosmetics retailer that used Scarfe's paint in its new warehouse and office building in Toronto. Scarfe recently shipped 200 five-gallon pails of recycled latex to Cuba, and ongoing talks with representatives of Peel Region in southern Ontario (where Laidlaw's facility happens to be located) about supplying all of that region's public buildings, as well as with several boards of education in Ontario, could result in a welcome boom for recycled paint. If so, Scarfe really will have to start to worry about sufficient supply. "The Peel Region contract alone could mean up to 300,000 gallons a year," says Coates.

But Coates still has his eye on our basements, which is why he's encouraged by suggestions of a Red Box program to go along with the current Blue Box system. The idea will be for homeowners to put their old paint cans in Red Boxes. Laidlaw's approach will be to pick them up on regular waste days along with the bottles, cans and newspapers they now collect. As Langdon points out, there are significant logistical problems with that—current regulations prohibit recycling liquids, for example, especially flammable, potentially toxic liquids such as paint—but if Red Boxes ever manage to get off the ground, Scarfe's supply problems will be over. "We all have cans of old paint in our basements," says Langdon.

The Greenhome team chose to use Scarfe's 100 percent recycled interior latex in the garage but decided against using it in the main area of the house. Recycled paint, because it is made from paint that may be as much as 10 years old, cannot be guaranteed to conform to today's standards for additives, so John was reluctant to use it in the house's living area. "We were going to go with a Glidden paint called Lifemaster 2000," says John, "but it's so chemical-free that it doesn't have any ethylene glycol in it." Virtually harmless unless you drink it, ethylene glycol, or ordinary antifreeze, is added to paint to make it conform to Canadian freeze-thaw standards; it also acts as a toughener. John chose an off-the-shelf Benjamin Moore interior latex, which has "just a little" ethylene glycol in it. "We figured that a little bit of toughener will mean that the walls will have to be painted less often, cutting down on the embodied energy. It's a trade-off, but trade-offs are necessary at some points."

□ □ □ □

HOUSES ARE BUILT TO LIVE IN AND NOT TO LOOK ON.
—FRANCIS BACON, 1612

Poor indoor air quality is a growing concern in all new houses and can be especially serious in airtight houses. It is caused by the release of toxic vapours from chemicals—binding agents, plasticizers, stabilizers, fillers, solvents and fungicides—that have been added to building, cleaning and decorating materials to make them stay wet longer or dry faster or resist decay and fungus. New materials made from hitherto unknown chemicals are now standard in new and renovated homes: foam insulations, polyethylene films, synthetic glues, caulkings and sealants. The effects of some of the volatile organic compounds (VOCs) that "off-gas" from these materials can be decidedly unhealthy. The number of people developing hypersensitivities to VOCs is increasing at an alarming rate, and part of the Greenhome's mandate to be gentle on the external environment includes creating a benign interior environment. In 1990, the Environmental Protection Agency in the United States put indoor air pollution at the top of a list of the 18 most dangerous potential causes of cancer—higher than low-level radiation, higher even than cigarette smoking. And the risk is greatest in new houses: studies comparing the concentration of organic gases and vapours in old and new houses found that the average in brand-new houses was more than 15 times higher than in houses more than three years old and, for some cancer-causing pollutants, as much as 100 times higher than the U.S. standard for *outdoor* air. "It is impossible to collect hard figures," John Maclennan, a clinical ecologist in Toronto, has said, "but there is no doubt that this is a big, big can of worms."

Not everyone is allergic to his or her house, but the American National Academy of Science recently determined that up to 20 percent of healthy adults exhibit some allergic reaction to formaldehyde, an organic compound found in just about everything that has glue or binding agents in it: insulation, adhesives, caulking, plastics, some wood products (the glue in plywood and fibreboard, for example), fabric finishes in carpets and drapes, foam rubber, even toilet paper and paper-towel rolls. According to Debra Lynn Dadd, author of *The Nontoxic Home* (1986), "formaldehyde is a suspected human carcinogen and has been shown to cause birth defects and genetic changes in bacteriological studies." She cites such symptoms as coughing, dizziness, headaches, rashes, tiredness, excessive thirst, nausea, nosebleeds, insomnia, disorientation and asthma attacks. Bruce Small, an engineer who has built a clean-air research centre north of Toronto and who wrote *Healthy Environments for Canadians* for Health and Welfare Canada in 1988, studied the air in several Toronto public schools and found 97 different sources of contaminants, including particleboard, shelving, desktops, carpet glue, felt-tip markers, photocopier chemicals, pesticides, cleaners and paint. "Some of the volatile chemicals found in construction materials and teaching and cleaning supplies," he wrote, "share characteristics with classic central-nervous-system depressants." Small believes that apart from the common symptoms of headaches and snuffles (the "summer-cold syndrome"), poor indoor air quality could be responsible for sluggishness, short-term memory disorders and learning disabilities.

Good ventilation is a band-aid answer to the buildup of toxic fumes in an airtight house—another reason, apart from moisture and odour problems, that mechanical ventilating systems are now mandatory in all R-2000 or tighter houses. But as Merilyn Simonds Mohr pointed out in "Air Apparent," an article on indoor air quality published in 1987, "the easiest way to improve indoor air is to make a clean start, choosing building materials that have few chemical additives and low emission rates." She quotes Bruce Small, who advises homeowners to "steer clear of things with a strong smell. The nose knows," he says. "If you want to test a substance for yourself, just take a deep whiff. If your eyes start to water, your nose runs and your sinuses ache, you know there is something in there that bothers you. We've generally found that if the smell is minimal, gassing-off probably is too." It's also worth pointing out that even so-called "pleasant" odours can be indications of excessive off-gassing of irritating chemicals—that new-carpet smell, for instance, could be the smell of formaldehyde.

The Greenhome design team made its interior decisions with indoor air quality in mind. It avoided vinyl plastics, for example, including polyvinyl chloride (PVC), one of the most dangerous plastics on the market. When PVC off-gasses, it releases vinyl chloride, which has been linked to cancer, birth defects and liver dysfunction. NASA has banned PVCs in space capsules because its off-gassing was so prolific, it gummed up the optical equipment. Most kitchen floors are made from it. So are some cosmetics, household cleaners, food packaging, crib bumpers and covers, pacifiers and teethers, inflatable swimming pools and plastic lunchpails. "Vinyl tiles are still the most common floor coverings in Canada," says Jeff Sellner, "at least in kitchens and bathrooms." The Green-home has ceramic tiles on the floors of the front airlock and the upstairs and downstairs bathrooms. The tiles, called Traffic Tile, are manufactured by the Stoneware Tile Company in Richmond, Arizona; they look like any other ceramic tile but are actually made of recycled glass. The adhesive and grout used to install them are both water-based products whose main ingredient is "fine quartz aggregate" – silica sand – mixed with a latex additive that acts as a binder and has, says Jeff, "virtually no off-gassing properties at all."

The rest of the floors on the second level, including the kitchen, are covered with reused hardwood. Reused hardwood makes good environmental sense: the Canada Mortgage and Housing Corporation has determined that at the time a house is built, the materials in it with the highest embodied energy are steel, cement and manufactured woods such as plywood; after the life-cycle costing over 40 years had been worked out, however, the product with the highest embodied energy is carpeting. Carpets are generally replaced every 5 to 10 years (closer to 10 in Waterloo, says Sellner, who sells them); hardwood floors will usually outlast the house and can often be salvaged and used again in another house. Sellner, who installed the hardwood in the Greenhome, says he has been getting more and more contracts for hardwood flooring in the past year or so. "It's becoming the most popular floor around," he says. "Even in kitchens. Mainly, I think, because it will last up to 70 years." Sellner got the hardwood in the Greenhome from Kieswetter Cartage and Excavation. "They're in the salvage business," says Sellner, using the old word for recycling. Kieswetter salvaged it from an old Seagram's Distillery warehouse just outside Waterloo, and Sellner's workers spent a day or so scraping solidified rye whisky out of the tongues and grooves before installing it in the Greenhome, sanding it and refinishing it with a water-based urethane (made by the Swedish firm Bona Kemi). When they were finished, it looked like a new floor.

To John, who won't drink coffee out of a Styrofoam cup, it looked better than new. "Of the three Rs," he says, "there's no doubt that re-using is better than recycling. There's a good reason that recycling is the third R." We're on our way to Waterloo's municipal waste-management facility, just beyond the Greenhome's subdivision. The back of John's Toyota station wagon is loaded with pieces of metal corner bead left by the drywallers and a bale of flattened cardboard from the boxes that the ceramic tiles came in. John had carefully picked up all the scrap and thrown it into the back of his car, and now we are taking it to the 1990s version of the local dump. In fact, that analogy has occurred to John.

"Last week I talked to the woman who manages the facility," he says, "and she told me that they sell the metal to a scrap dealer in Hamilton, and the cardboard goes to Domtar, who makes new cardboard out of it. I've even heard that they can make new white paper out of recycled cardboard using a bleachless process, which would be nice. But I don't know if that's real or just greenwashing. I suppose there are some things we just have to take on faith. Still," he adds, "I think the real answer, the most environmentally sound answer, is to reduce. I mean, we're talking about *excess* here. We *use* too much stuff. What are the three Rs? Reduce, reuse and recycle. And the greatest of these is reduce."

In 1991, Ontario's energy minister, Ruth Grier, established something called the Waste Reduction Office (WRO), a government agency whose job it is to institute a Waste Reduction Action Plan (WRAP) that would make Blue Boxes mandatory throughout the province, conduct waste audits on manufacturers and municipalities and force industries, institutions and shopping malls to separate their solid waste and recycle anything recyclable. Other aims of the plan developed when Ontario's Waste Management Act came into effect in April 1992: to divert 25 percent of Ontario's solid waste by the end of 1992 and 50 percent by 1997; to prohibit the long-distance transportation of solid waste out of Toronto; and to ban the construction of waste incinerators. Grier also announced that an astounding 127 of the province's 220 landfill sites would be closed by February 1997. In other words, the WRO was going to force the province to reduce the amount of waste it produced by gradually eliminating the places it could take the waste to. When I'd called the WRO a few days earlier to find out how much household solid waste was being recycled in Ontario, a woman at the office said they didn't have that statistic right at hand. "This is the Waste *Reduction* Office," she said pointedly. "You're asking for a waste *diversion* figure." I hadn't really considered the distinction before. John, as usual, had.

"The best way to deal with excess," he is saying, "is to not create it in the first place. Take paint. I'm not so sure that you don't use as much energy to break down old paint and make something new out of it as you do to manufacture new paint. It seems to me to make more sense to not have any old paint left; just order the exact amount you need, and use it all up. If you have any paint left over, put another coat on the wall. Just think how much all this recycling is costing us."

As we pull into the waste-management depot, a huge sign beside the toll gate tells us exactly how much recycling is costing us. Dumping up to 50 kilograms of anything is free, but after that, it begins to cost. Material that could go into a Blue Box is relatively inexpensive: from $1 for

51 kilos to \$20 for more than 200 kilos. The most expensive item is rubber tires: \$1.50 each or \$159 for more than 200 kilos. You can dump a dishwasher here for about \$13.50 or a refrigerator for \$45.

Vehicles are weighed on the way into the yard, weighed again on the way out and charged for the difference. When John pulls onto the scales at the toll gate, the woman asks him what he has in the back. Then: "Steel in Bin 14, cardboard in Bin 8," she says. "Will that be cash?"

CANADA IS CONFRONTING A MAJOR WATER CRISIS, A
SHORTAGE THAT COULD LEAVE NORTH AMERICA
UNRECOGNIZABLE. CANADIANS HAVE BEEN SO LULLED BY
THEIR OWN MYTHOLOGY, THEIR VISION OF A LAND OF
COUNTLESS LAKES AND MYRIAD UNTAMED RIVERS, THAT
THEY HAVE NOT YET NOTICED THE DANGER, LET ALONE
STARTED TO AVERT IT.
—JOHN SEWELL, QUOTED IN *THE BROWNING OF AMERICA*,
BY SUZANNE ZWARUN, 1982

[TOILETS ARE] THE BIGGEST WASTE OF WATER IN THE
COUNTRY BY FAR: YOU SPEND HALF A PINT AND FLUSH TWO
GALLONS.
—THE DUKE OF EDINBURGH, IN A SPEECH DELIVERED
IN 1965

The average human being in North America generates 130 gallons of body waste per year. The average toilet in North America uses 5 gallons of water per flush, which means that each of us uses 9,000 gallons of water a year—about half our total domestic use of potable water—to get rid of 130 gallons of waste, a ratio of about 70 to 1. It seems rather obsessive, doesn't it? It would appear that our approach to water management, as described by Sir Thomas Crapper's compatriot the Duke of Edinburgh, has helped to create the crisis in water supply lamented by former Toronto mayor John Sewell.

One-quarter of the world's supply of fresh surface water lies within Canada's borders, and at one time, 30 times that amount lay under its surface. Today, a good percentage of our surface water is unusable unless treated (the Great Lakes system from Sault Ste. Marie to Cornwall was first declared "polluted to such an extent which renders the water in its unpurified state unfit for drinking purposes" in 1912), and no one

180

knows for sure how much groundwater is left in our aquifers. In the Kitchener-Waterloo area, the indications are that it isn't much.

The disappearance of our groundwater is a recent phenomenon. When Daniel and Jacob Erb gave Richard and Henrietta Beasley £10,000 for Block 2 on the Grand River in 1805, one of the area's chief attractions was its plentiful supply of groundwater: by 1817, when the name Waterloo was registered, the region was famous for its artesian wells. One hundred and fifty years later, the water situation was still comfortable: an engineer's report in 1966 declared that the city was using only 13 million of the 22 million gallons available to it each year. Then something happened. In the early 1970s, massive population growth increased the area's annual water demand to nearly 40 million gallons, and at the same time, the boom in construction reduced the amount of rainwater that percolated down into the ground—more than two-thirds of Ontario's Golden Horseshoe, the industrial curve that wraps around the end of Lake Ontario, is under pavement or roof—and local wells began to run dry. In 1977, the chairman of the Waterloo Regional Council announced that given current rates of consumption, Waterloo's supply of fresh water "could not be proven beyond 1988."

There ensued a spate of the kinds of proposals that have become so typical on this continent: when faced with serious long-term shortages—whether of oil or electricity or fresh water—instead of cutting down our demand, we look around for ways to increase our supply. The Ontario government of the day, through its Water Resources Commission, proposed to build a water pipeline to Kitchener-Waterloo from Nanticoke, a town on Lake Erie where a huge steel mill was already sucking vast quantities of water out of the lake (it takes 40,000 gallons of water to produce a ton of steel). Kitchener pooh-poohed the proposal, preferring instead to run a pipeline into Lake Ontario because it foresaw more opportunities for industrial growth—which meant even more water use—along the proposed right-of-way. Local residents, on the other hand, said they would rather have no pipeline at all, but if one proved necessary, they would like it to bring in water from Lake Huron. The town of Brantford suggested damming up the Grand River and holding the water in a huge reservoir. A study was commissioned to weigh the relative merits of each proposal; the Lake Erie pipeline (which will cost $230 million) is now running slightly ahead of the Grand River dam (which would cost only $7 million); in the meantime, water is being piped out of the Grand River at nearby Mannheim, treated and either used right away or stored in the depleted underground aquifer, where it may or may not be hanging around waiting to be pumped back out again when necessary.

DAILY WATER USE

	WATER USE (litres/day)	
	OBC* HOUSE	GREENHOME
Lawn	96	0
Sinks	34	21
Toilet	260	39
Bathing	164	115
Appliances	96	42
Water Treatment	34	11
TOTAL	683	228

*Ontario Building Code

Needless to say, interest in water conservation is keen in the Kitchener-Waterloo area. Studies in the United States have shown that when it comes to water conservation, the stick works better than the carrot: when Elmhurst, Illinois, passed around low-flow shower heads and plastic displacement bottles to put in toilet tanks, its water consumption rate went down 15 percent; a similar experiment in Pennsylvania resulted in a decrease of only 10 percent. In programs in Maryland and Virginia, however, in which the low-flow devices were actually installed in apartment buildings by the Metropolitan Watersaving Company instead of being simply handed out, the savings were considerably higher: 22 percent in Maryland and 40 percent in Virginia.

It was no accident that all these water-saving measures were aimed at domestic toilets and bathtubs: depending on the area, domestic water use accounts for about half of a community's total consumption, and half the water each of us uses every day goes down one or the other of those drains. It is another of life's supreme ironies that the first toilet — patented in 1882 in England by the unfortunately named Sir Thomas Crapper — was called the Crapper Valveless Water-Waste Preventer. (Although it may be true that Sir Thomas lent his surname to the device, it is definitely not true that it became a verb on his account. The *Oxford English Dictionary* gives the first use of "crap" as "to defecate" in 1846; Sir Thomas must have had a very difficult time of it in school, which may have implanted in him the idea for his invention.) Most toilets today are far from water-waste preventers. Although homebuilders everywhere routinely install toilets with at least 3.5-gallon flushes but usually 5-gallon flushes, studies conducted in Phoenix, Arizona, as long ago as 1986 showed that ultralow-volume (ULV) toilets work just as well as and sometimes better than conventional toilets at removing waste and save up to 30 percent of the house's overall water usage while doing it.

The study compared two Phoenix subdivisions, identical in every way except that one had conventional toilets with 3.5-gallon flushes and the other had ULV toilets with 1.5-gallon flushes. The study set out to determine whether ULV toilets would work as well as standard toilets—conventional wisdom had it that low-flush toilets would have to be flushed more often and also would not be as clean as regular toilets—but what it discovered surprised even ULV supporters. At the end of the year-long study, the researchers found that conventional toilets had required double-flushing more often than the ULV toilets and had clogged more often than the ULV toilets and that the subdivision with conventional toilets had had more sewer-line blockages than its ULV counterpart. Did that study change anyone's mind about what kind of toilets to install in new subdivisions?

Well, not universally. But things might be changing nonetheless. There are carrots: Glendale, Arizona, for example, offers $100 to every person who replaces a conventional toilet with a low-flush model. And there are sticks: Morro Bay, California, has instituted a bylaw whereby builders receive permits for new buildings only after they have saved as much water as the new buildings will use. They can either pay to have some of the city's old water pipes replaced or pay to install low-flush toilets and low-flow shower heads in 12 existing buildings.

Organizers of a recent conference on water management in Winnipeg, Manitoba, asked Ian Cook, as president of the Kitchener-Waterloo Home Builders Association, to give a talk on water management from a homebuilder's perspective. "What the heck do I know about water management?" he asked John. But when he and John sat down and looked at some numbers, Ian realized he knew a lot about it. For one thing, he knew that the city of Waterloo had recently begun passing on the cost of expanded water services to developers, increasing development fees to offset the extra pressure new subdivisions were putting on the city's water supply and treatment facilities; if a developer could lower a subdivision's water use, then the development fees would go down. A pilot project conducted in Kitchener in 1991, in which the 5-gallon-flush toilets in 500 houses had been replaced with 1.5-gallon-flush toilets, had resulted in a 30 percent reduction in overall water use. Ian calculated that installing a ULV toilet in an average house would save the homeowners $90 a year in water bills and installing ULV toilets in a 200-unit subdivision would save the developer thousands of dollars in development

GREENHOME WATER-EFFICIENCY MEASURES

THE BATHROOM – Application of high-efficiency technologies could reduce water use in the bathroom by 75% or more

TOILETS: Litres per flush

CONVENTIONAL	MIDEFFICIENCY	HIGH-EFFICIENCY	GREENHOME CHOICE	POTENTIAL SAVINGS
14 to 23 L/flush	13.5 L/flush	2 to 6 L/flush	Fluidizer: 2 L/flush	Annual savings for a family of four: 132,000 L

FAUCETS: Litres per minute

CONVENTIONAL	MIDEFFICIENCY	HIGH-EFFICIENCY	GREENHOME CHOICE	POTENTIAL SAVINGS
Average flow rate: 13.5 L/minute	Reduces flow rate to 6 to 9 L/minute	Reduces flow rate to around 2 L/minute	The Omni AS-4020, with a flow rate of 2 L/minute	A low-flow faucet reduces water use by 11.5 L/minute, a reduction of 85%

SHOWERS: Litres per minute

CONVENTIONAL	FLOW RESTRICTORS	WATER EFFICIENCY	GREENHOME CHOICE	POTENTIAL SAVINGS
Flow rates up to 20 L/minute	Specially designed washers reduce flow through showerhead	Low-flow showerheads have flow rates between 6 and 10 L/minute	The Ondine Super Watersaver #29446 at 7.6 L/minute	The Ondine unit reduces consumption by over 60%, an annual savings of just over 90,000 L for a family of four

THE KITCHEN – Application of high-efficiency technologies could reduce water use in the kitchen by up to 25%

DISHWASHERS

CONVENTIONAL	WATER-EFFICIENT		GREENHOME CHOICE	POTENTIAL SAVINGS
Up to 1,200 L per month	Few water-efficient dishwashers are readily available		Miele Turbothermic G 590 SC	Using a dishwasher only when it is filled to capacity saves several operating cycles per week

FAUCETS: Litres per minute

CONVENTIONAL	MIDEFFICIENCY	HIGH-EFFICIENCY	GREENHOME CHOICE	POTENTIAL SAVINGS
Average flow rate: 13.5 L/minute	Faucets that reduce flow rates to 6 to 9 L/minute	Faucets that reduce flow rates to about 2 L/minute	Higher flow rates are needed in kitchens for filling pots and sinks. Recommended: the Vandenburgh model SSOA, at 7.6 L/minute	A midefficiency faucet can reduce annual water use in the kitchen by about 25%, or 6,500 L

THE LAUNDRY-UTILITY ROOM

CLOTHES WASHER

CONVENTIONAL	WATER-EFFICIENT		GREENHOME CHOICE	POTENTIAL SAVINGS
A clothes washer uses 150 to 250 L of water per cycle—1,200 L annually for a family of four	European appliances tend to be more water-efficient; front-loaders use up to one-third less water		Miele Novotronic W 1918; look for automatic washers that have a suds-saver, as well as variable load settings; appliances with low EnerGuide ratings should be given preference	Using a clothes washer only when it is filled to capacity saves several operating cycles per week

HOT-WATER TANK

			GREENHOME CHOICE	POTENTIAL SAVINGS
			Insulating the tank and hot-water pipes will reduce standby heat loss	Water-heating costs can be reduced by as much as 25%

185

fees. "I think the other builders were impressed," he told John when he returned from Winnipeg. Ian was a bit impressed himself. So was John. He went out and bought three ULV toilets and replaced the 3.5-gallon toilets in his own house. "They cost only $30 apiece more than conventional toilets," he says. "They'll pay for themselves in five years."

The toilets in the Greenhome are even more impressive. The toilet in the downstairs bathroom is made by Ifö Sanitär, a Swedish company whose "water closets" use only ⅘ gallon of water per flush. The toilet looks more or less like a conventional one, except that its tank is somewhat rounded and the tank top looks like the lid of a teapot. It also behaves like a conventional toilet: it works by a gravity flush and achieves its purpose by splitting the tank discharge into two portions, one on either side of the bowl, which creates a swirling action that makes better use of the water.

If both toilets in the Greenhome were of this type, it would mean water savings of more than 5,000 gallons a year for an average-size family. If the conventional toilets in every house in North America were replaced with ULV toilets, we would save almost 2 billion gallons of fresh, drinkable water *every day*. Although that represents less than 1 percent of our total water use, when we consider how much money we spend on water treatment, desalination and chlorination as well as the environmental costs of water storage and transportation, new dams and reservoirs, it adds up to significant savings.

The toilet in the upstairs bathroom, however, uses even less water than the Ifö. It is called the Fluidizer and is made by Control Fluidics, a company based in Greenwich, Connecticut, a city that knows a lot about water conservation. During a major drought in 1980 and 1981, the citizens of Greenwich cut their water consumption by 25 percent by such commonsense measures as washing their cars less often, taking fewer baths and not leaving the taps running while they brushed their teeth. If they had all installed Fluidizer toilets, they would have reduced their consumption by more than 50 percent, because the Fluidizer uses only ⅖ gallon of water per flush, which represents a 90 percent saving over conventional-flush toilets.

It accomplishes this marvel by means of something Control Fluidics calls "hydraulic attrition" combined with "velocity jet rinse," which means that the toilet liquidizes the solid waste deposited in it, including paper, by pulverizing it with a jet of water for eight seconds. "It's sort of a mini-garbage disposal unit," says John. It is hooked up directly to the water lines, so it does not require a tank, cannot condense or leak (approximately 60 percent of all standard toilets leak water constantly, a further waste) and can be reflushed immediately. There are other manufacturers of minimal-flush electronic toilets – a close contender for the Greenhome was one called the Microflush Trimline, made by Microphor Inc. in Willits, California. Both the Microflush and the Fluidizer use about 2 litres of water per flush and cost about the same ($495 U.S.), but the Microflush has a compressor in it that uses 20 kWh of electricity per person per year, whereas the Fluidizer has no compressor and uses only 1.4 kWh per person per year. "The difference isn't amazing," says John. "I mean, we're talking about maybe $1.50 a year differential in electricity costs. But we had to pick one, and we had to base our decision on something, so we went with the lower embodied energy of the Fluidizer."

It should also be remembered that neither toilet in the Greenhome is connected to the house's drinking-water supply; both are hooked up to the cistern, which recycles rainwater, so the Greenhome is already eliminating up to 40 percent of the potable water used in a conventional house. Not every water-using device can draw water from the cistern, however, so other conservation measures come into play. After toilets, the second-largest water users in a house are the bathtubs and showers: about 25 percent of our household water goes down a tub or stall drain, and most of it is hot water, which adds to the energy component. A full bathtub uses 60 gallons of water; showers use less, but not much less, depending on how long you take to have a shower; conventional shower heads have flow rates of about 5 gallons per minute, so if you stand under one for 12 minutes, you might as well be taking a bath. Many low-flow shower heads are available; these work like aerators on kitchen taps, by mixing air with the water so that the pressure feels the same but the volume of water is reduced. The Greenhome's two showers are made by Moen and use only 1.8 gallons of water per minute. John calculates that a typical family of four taking five-minute showers every day will use about 40,000 gallons of water a year; with the Moen units, that figure will drop to 13,140 gallons. "The new Ontario Plumbing Code is going to *require* shower heads with a flow rate of 1.8 gallons per minute," says John, "so a lot of companies are going to start making them."

The Code is also going to stipulate that faucets have flow rates of less than 2 gallons per minute. Conventional faucets allow water to flow at about 3.5 gallons per minute; high-efficiency faucets are available that cut the flow to half a gallon, "but that's too slow, I think," says John. "At that rate, it would take 10 minutes to fill a sink for doing the dishes. People won't wait that long; they'll get fed up and just take the aerators out." The faucet in the bathroom, made by Nepitek in Nepean, Ontario, comes equipped with an electronic sensor that turns the water on automatically when an object – say, a toothbrush – is placed under it and off when the object is removed. This is more than a Star Trek gadget: thousands of gallons of water are wasted daily because we leave the tap running when we're doing something else. Brushing our teeth takes about three minutes; if a conventional faucet is left open even for that long, 10.5 gallons of water run down the drain. Do we need 10.5 gallons of water to brush our teeth? The arithmetic isn't difficult: even at the low rate North American householders are taxed for their water (42 cents per 265 gallons) compared with Europeans (86 cents) or Australians ($1.47), an average family spends twice as much money on water to brush their teeth each year than they do on toothpaste.

□ □ □ □ □ □

NO HOUSE SHOULD EVER BE *ON* ANY HILL, OR ON
ANYTHING. IT SHOULD BE *OF* THE HILL, BELONGING TO IT,
SO HILL AND HOUSE COULD LIVE TOGETHER, EACH THE
HAPPIER FOR THE OTHER.
— FRANK LLOYD WRIGHT, *AN AUTOBIOGRAPHY*, 1932

"The architecture of the Americas," Arthur Erickson told the American Institute of Architects in 1988, "is heir to the self-obsessed Mediterranean world. In our reckless conquest of the land, we stifled the native Oriental spirit and proceeded with defiant ruthlessness to clear and build." We cleared and built at such a violent pace, he said, that we destroyed the natural landscape of our continent and ended up with "the bristling towers and desolate streets of our urban cores and the asphalt sprawl of the suburbs." This has led to a spirit of self-reliance, of practicality and inventiveness, but it has also produced "our main neurosis: our guilt for our abuse of nature."

Don Prosser doesn't think Erickson overstated his case. As the landscape architect charged with designing the Greenhome's yard, Don was very aware of the need to take the Greenhome's internal philosophy and apply it to the outside of the house. At the design meetings, he spoke of "landscaping as atonement," of "giving back to nature what our bulldozers have taken away from it." He introduced the concept of "the entire environment, as opposed to the enclosed environment."

The subdivision site plan already provided a list of plants, trees and shrubs that would be put in place by the developer. The list included sprinklings of honeysuckle, Chanticleer pear trees, alpine currant bushes, forsythia and Saskatoon berries as well as the more obvious mountain ash, white spruce and sugar maple saplings. But the main approach to landscape design for Freure, as for most developers, was still: Where do we put the sod?

"I think turf has had a bad rap," says Don, using the landscape architect's term for grass. After all, landscape designers must have the same sort of fondness for grass that landscape painters have for tubes of chrome green paint – it covers a lot of otherwise problematical territory. "Turf is a good ground cover," Don says. "It withstands foot traffic better than any other ground cover I know. But," he admits with a tinge of regret, as though talking about a family member that has gone wrong, "it's a high-energy ground cover, to say the least. It needs a lot of attention. It needs a lot of cutting. And it needs a Niagara of water."

North Americans lavish millions of gallons of water on their lawns every year: typical lawn sprinklers "sprinkle" 8 gallons of water per minute, and on hot days, one-third of that evaporates before it ever hits a blade of grass. And even in areas of high annual rainfall, the preponderance of grass as a ground cover seriously impedes the passage of rainwater into groundwater. To Malcolm Wells, grass is just another form of pavement. "Each year," he writes in *Gentle Architecture*, "we draw another 10 million acres from the green side of the national ledger. As those acres are turned into what we call improved land, they become very efficiently paved, if not with blacktop or concrete or roofing materials, then paved with closely mowed turf – lawn grass – which is no slouch as a paving material either. Neatly trimmed grass can be counted on to repel almost half as much rainwater as a shingled roof." Instead of soaking into the earth and filtering into our aquifers, as it would if it fell on undeveloped rivers, away from the land. It is not particularly surprising, then, that our aquifers are drying up.

So Don Prosser chose other plants for the three areas on the Greenhome lot designated for ground cover. Ground A, as the area in front of the house between the sidewalk and street is called, "is an excellent place for buffalo grass." Buffalo grass is a native species (unlike the Kentucky bluegrass, red fescue and perennial ryegrass that make up most Canadian lawns) that is extremely drought-resistant – in fact, it never needs to be watered at all – grows to about 3 inches in height and then stops, so it doesn't need to be mowed either, "and it's very tough," says Don. "You can play football on it." If buffalo grass has a drawback, it's that it browns out at first frost, which in Waterloo is usually in September. Don doesn't see that as a huge problem: "We have to learn to walk the line between aesthetics and responsibility to the environment," he says, "and that line may very well be sown with buffalo grass." 189

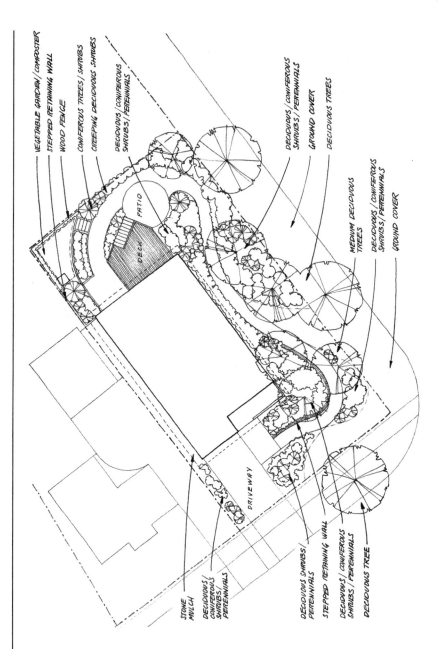

VEGETABLE GARDEN / COMPOSTER

STEPPED RETAINING WALL

WOOD FENCE

CONIFEROUS TREES / SHRUBS

CREEPING DECIDUOUS SHRUBS

DECIDUOUS / CONIFEROUS SHRUBS / PERENNIALS

DECIDUOUS / CONIFEROUS SHRUBS / PERENNIALS

GROUND COVER

DECIDUOUS TREES

MEDIUM DECIDUOUS TREES

DECIDUOUS / CONIFEROUS SHRUBS / PERENNIALS

GROUND COVER

PATIO

DECK

DRIVEWAY

STONE MULCH

DECIDUOUS / CONIFEROUS SHRUBS / PERENNIALS

DECIDUOUS SHRUBS / PERENNIALS

STEPPED RETAINING WALL

DECIDUOUS / CONIFEROUS SHRUBS / PERENNIALS

DECIDUOUS TREE

The Greenhome's lawns are sown with buffalo grass and
white clover, which require no watering or mowing. Plants
and shrubs are native to the area. Some are edible; others are
evergreen, providing visual interest year-round. Trees are
placed so as not to block winter sunlight from the windows.

Ground B, along the Bridgewater curb, is sown with white clover. "White clover is also extremely drought-resistant, and it doesn't need to be mowed either, as long as you can accept a ground cover that's a foot high." Don says that white clover will undergo a little winterkill but always fills in nicely in the spring, and in late summer, it will put out small white flowers that attract butterflies and bees. About halfway along the southern edge of the property, a tall green ash tree marks the beginning of Ground C. Don hesitated for a while over this area. He considered periwinkle because he likes its brisk, shiny leaves, but in the end, he chose *Euonymus fortunei*, "a leathery sort of broadleaf evergreen that keeps 80 percent of its leaves over the winter, when they acquire a somewhat reddish hue." As this area is directly in line with the Greenhome's living room windows, the winter colour was thought to be a definite plus.

Technically speaking, the three ground-cover areas were all outside the Greenhome's lot line; within it, Don had to contend with several permanent fixtures that to some extent, dictated the shape of his landscape design. First there is the driveway, which is paved with an open fretwork of turfstones, the holes of which are filled with topsoil and sown with buffalo grass. Then there's a turfstone wheelchair-access path that runs from the driveway, across the front of the house and along the side, then curves around the back and down into the lower-level walk-out. Just where the path curves at the back is a turfstone patio that leads to a triple-layered natural cedar deck that climbs up to the second-floor walk-out. Don prepared the ground on either side of this path by tilling down through a foot of topsoil to the heavy clay subsoil and mixing in a gypsum loosener made from recycled drywall. "It gives excellent drainage," he says, "and it doesn't break down as fast as organic looseners." Also mixed in with the topsoil is an organic compost that he gets from a nearby mushroom farm; the mushroom farm gets it from a trucking firm that hauls it from a nearby horse barn. "I had to determine that the mushroom farm didn't use any chemical fertilizers," he says. "Sometimes it has a high salt content, but when you till about 2 inches of it into 12 inches of soil, the salt seems to be just about right."

When the topsoil was prepared, Don put in the plants. His preference was for native varieties. "I looked for plants native to this region first, then to southwestern Ontario in general, then to Ontario. We've threatened or destroyed so many of our native species over the years," he says, "that I think we owe it to the land to put some of them back."

His second guiding principle was: No annuals. "Annuals are too energy-consuming," he says. "They're an indoor crop, essentially: they're started in the greenhouse, left in the greenhouse much longer than perennials, and it takes a lot of energy to keep a greenhouse heated. Plus, when they're in the ground, annuals require a lot more water than perennials. And then you have to go through it all again a year later. Perennials are valued for their foliage as much as for their flowers, and they can be kept going for a long time. Some peonies, for example, can last 25 or 30 years."

The bed between the path and the Greenhome's south-facing wall is planted in hostas; facing them across the path are rows of viburnum (high-bush cranberries) and serviceberries—"Food for people as well as birds," says Don. He also put in such natives as St. John's wort and flowering quince, a shrub that grows to about 5 feet, whose fruit the early settlers used to make jam; some low evergreen junipers whose blue berries have been used to flavour gin but which taste almost as wonderful in stews; and day lilies, the roots and young flower heads of which can be eaten. Under the American yew and dogwood trees that grace the yard's corners are the shade-tolerant Solomon's seal, lady fern and bearberry bushes, whose broad evergreen leaves also turn red in the fall. "People tend to think of their yard as a pleasant thing to look at in the summer," says Don. "But, especially in Canada, it's important to consider what it will look like the other 10 months of the year."

When I left Don's office at York Nurseries, just outside Waterloo, it was mid-March. Although the sun was beginning to press against the skin with the warm purr of spring and the air seemed warmer than it had that day in September when we began excavating the Greenhome's foundation, there was still snow on the ground. A brisk wind was making little snow dunes against my car tires. The radio was issuing warnings of the worst storm of the century blowing its way up the Atlantic Seaboard from the Gulf of Mexico, threatening 5 feet of snow in the Adirondacks. I drove over to the Greenhome. Inside, the fireplace was lit and the house was warm. Looking out through one of the south windows, I imagined banks of red-leafed bearberry bushes under an intricate pattern of snow-laden mountain ash branches. Two bare honey locusts broke my view of the street without impeding the flow of winter sunlight into the room. A Morse code of rabbit tracks crossed the yard, stopping at serviceberry and cranberry stalks, then continuing to the back to check out the compost pile beside the vegetable garden. The rabbit's movement triggered the switch on the solar-powered yard lamp, which cast a soft glow of free light across the sparkling undulations of snow. Yes, I thought, this house can work.

OPENING

□

MAN SEEKETH IN SOCIETY COMFORT, USE AND PROTECTION.
— FRANCIS BACON, *THE ADVANCEMENT OF LEARNING*,
1605

The first time I see 302 Westvale Drive in its finished form is just after noon on April 25, the day the Greenhome is opened to the public. Westvale Drive has been blocked to traffic, and tents have been set up from which Union Gas employees are distributing free hot dogs and soft drinks to the crowd who has come to see what the *Kitchener-Waterloo Record* is calling "a showcase of environmental design and technology that is full of bright ideas you can use in your own home." Beyond the tents, a van painted to look like a radio blasts out a string of golden oldies to a long line of people standing patiently in the rain at the Greenhome's garage door, waiting for another long line of people to finish touring the house and make room for them. I find John and Steve in the garage, surrounded by people taking off or putting on shoes: "We were told to expect about 2,000 people. We've had nearly 700 so far," says John happily. "Looks like we're going to make it."

Other members of the Greenhome team, each wearing Greenhome windbreakers and T-shirts donated by Union Gas, are scattered about the house and grounds, directing traffic, talking about the house's low-energy features, munching on hot dogs, looking busy.

"I love the layout," someone says to architect Richard Reichard. "It looks small from the outside, but when you get in, it's quite spacious. It makes very intelligent use of the space."

"So many people were involved in the planning and design," Richard replies modestly, "that I hardly feel responsible for it. But you're right, the idea was to make every square foot count. People appreciate common sense these days, especially here in Waterloo. I think it's the German background."

Elizabeth White is talking to Tony Krimmer about the cistern. "How did you get approval for two separate plumbing systems, Tony?" she asks. "Does the Code approve of one system for potable water and another for nonpotable water?"

Tony shrugs. "Sure. There's nothing in the Code that says it can't be done. We had to submit drawings to show that the two kinds of water couldn't get mixed anywhere, but once we did that, there was no problem."

Elizabeth tells him about a recent case in Belleville, Ontario, in which

194

a builder had tried to get permission to install a cistern and use the water to flush the toilets in the house. "They told him it couldn't be done," she says. "The Code wouldn't allow it."

"That's just a lot of noise," says Tony. "The Code is just people. Tell him to keep talking. That's one of the great things about the Greenhome," he says. "It's going to open a lot of people's eyes about what can and can't be done."

John has moved into the dining room and is explaining the dormer windows above his head to a group of visitors. He has put plaques on the walls pointing out some of the house's innovative features, and the one under the dormer window reads: "Automatic Shading. A 'smart' weather panel over the dormer windows and centre dining room window contains a gel which becomes opaque and reflects the sunlight once the indoor temperature reaches 24 degrees C. This reduces the cooling requirements during the summer."

"How does the gel work?" a woman asks him, looking up into the vaulted ceiling.

"Well," says John, "it's a thermochromatic film that fits over the glass, and it works just like the photochromatic film on some eyeglasses, the kind that turn into sunglasses when they are worn outdoors. Only, with these windows, it's heat that makes them become opaque, not light. When the surface temperature of the glass hits 24 degrees C, the film clouds up and blocks any more sunlight from coming in. If it still gets too hot in the house, we can open the dormer windows by flicking this switch" – he reaches over to a switch on the dining room wall and flicks it. Twenty feet above his head, both dormer windows whir and open.

"I don't know about all this environmental stuff," a man beside him says, "but I sure do like this house."

In the kitchen, more people are checking out the expensive-looking European appliances, mentally subtracting monthly hydro bills from the added purchase price and looking thoughtful. Above the white countertop, another one of John's plaques explains that "a teakettle is the most efficient way to electrically heat water. It uses less electricity and is faster. The thermal-carafe coffee maker is also very efficient, because the coffee stays warm once it is brewed without having the heating element on constantly. The range uses a sealed ceramic countertop which eliminates open flames in the home. Combustion air is brought in from outside, and combustion products are vented to the outside through the wall. Burning natural gas in the house releases one-sixth the amount of carbon dioxide as using electricity from a coal-fired power plant."

I join a touring group on its way downstairs and jot down a few of the things that have been put in at the last minute. The floor of the cold cellar in the basement, for instance, has been covered with tiles made from recycled sheet plastic. John had taken all the low-density polyethylene that had wrapped the Greenhome's insulation and other building materials to Terra Plastic in Cambridge, and Terra had given him a stack of tiles in return. "They just melt it down and pour it into moulds," says John. "Look, you can still see some of the printing in them."

Another feature I haven't seen before is the glass window in the partition wall between the upstairs bathroom and the great room: Tom Uznanski, a local craftsperson, fitted panes of glass about a half-inch apart in a wooden frame, then filled the space between the panes with crushed clear glass, or "flint," from a recycling centre. The resulting wall is a statement: environmentalism as art. It seems a perfect complement to the statement made by the whole Greenhome: it is aesthetically pleasing (it looks winter-frosted, crystalline), it is practical (it allows light from the dining room dormers to penetrate deep into the bathroom on the north side of the house), and it is made from recycled material.

By five o'clock, the predicted 2,000 people have seen the Greenhome and have left, and all is quiet. John stays behind for a few minutes before turning the keys almost reluctantly over to me, since I am going to stay in the house overnight. He seems tired but elated. "John has been living and breathing this house for months," says his wife Maggie when she comes to pick him up. "When he started working at Enermodal, he was doing hands-on engineering with heat exchangers, and he loved it. Then he started spending more time in the office, wearing ties and jackets, and I think he was looking for more hands-on work again. He certainly found it in the Greenhome."

When they leave, I make a second, private round of the house. I go down into the fruit cellar, where the precast concrete foundation wall is still visible, and remember that warm week in September when the panels were craned into place and I helped Peter and Bert bolt them together. I also remember the day Phil was fixing the sill plate to the top of the basement wall and I was standing below him looking at something on the ground, when a 12-inch bolt slipped from his hands and hit me on the back of the head just below my hard hat. From the fruit cellar, I walk back through the downstairs bathroom, along the hall and into the mechanical room, a maze of wires, pipes, ducts and meters. The furnace/HRV is humming away as I enter, and a small whine from one end of it tells me that the ventilator is drawing exhaust air from the kitchen upstairs and the two bathrooms.

A display in the downstairs family room lays out most of the Greenhome's more tangible energy-saving features, with samples of materials, and I walk along the table making note of them. There is a jar of recycled cement, crushed into aggregate, that was used in the weeping tile bed; a piece of yellow fibreglass batting made from recycled glass that insulates the garage wall; a square of wallboard made from recycled newsprint, harder than drywall and without the paper backing; a section of a wall joist made from recycled wood fibres. The list is familiar to me, yet still strange. Even after a year of watching, reading, talking and writing, I am amazed at the potential for lowering a house's impact on the environment. "We can't eliminate the impact," Ian Cook said earlier that day, "but we can reduce it to a minimum. So much of what we do is out of habit, unquestioned. All the Greenhome does is ask why."

Why. Why keep mining new bauxite when there is so much aluminum already made that we could be using? Why turn radioactive energy into electricity to heat our houses when every other fuel source is cleaner, more efficient, less expensive and less dangerous? Why build basement walls with twice as much concrete as is needed? Why spend millions of dollars purifying our water so that we can drink it safely and then use almost half of it to flush our toilets? Does it make sense? Does it make more sense to do it another way?

The Greenhome has proved that answering these questions does not condemn us to living in unattractive modules, uncomfortable cells or unaffordable mansions. As I sit in the early evening sunlight that streams across the hardwood floor (the rain has stopped, and the sun is struggling to make its own environmental statement), the wall thermostat registers a comfortable 72 degrees F even though it is set at 67. The air, even after 2,000 hot-dog-carrying visitors, is fresh and clean. I think again: yes, this house works. It is an attractive and intelligent blend of practicality and aesthetics. The foot-thick walls keep the house warm and quiet and also provide wide window ledges for houseplants, sills as deep and comforting as the window recesses in stone farmhouses. The compact fluorescent lights that I now have to turn on with the onset of twilight also perform a dual purpose, lighting the entire room with just 26 watts of electricity, yet providing plenty of warm, soft, yellow light for me to read by: I have brought Witold Rybczynski's *Home* along for inspiration. Even the furniture is a perfect marriage of form and function: donated by a local antique dealer, the dining table and chairs, the desk in the corner of the living room, the side table in the bedroom were all handmade a century ago out of solid wood, with no modern chemicals, no laminated particleboard, no formaldehyde glues, and they glow in the soft light with a quiet, firm dignity, perfect examples of older materials pressed back into service. Even the downstairs bathtub is one of those claw-footed enamelled tubs, refinished, restored and ready for reuse.

In fact, filling the house with 100-year-old furniture is appropriate on several levels. In a way, the Greenhome represents a return to such 19th-century concepts as common sense, efficiency and comfort. Unlike many modern, even low-energy houses, the Greenhome is not high-tech; it is determinedly low-tech. I have lived in turn-of-the-century farmhouses, in which many of the "bright new ideas" featured in the Greenhome were still in everyday use: the cistern, the metal roofing, wood flooring, curtains, upholstery and bed coverings made of real cotton, passive solar heating, rag-washed-and-striped painted walls instead of vinyl coatings and fast-drying glues. Even old newspapers used for insulation, come to think of it. The houses weren't as convenient as the Greenhome, perhaps, but convenience is not the only component of comfort.

"What is comfort?" asks Witold Rybczynski in the final chapter of *Home*. It is not, he concludes, a simple matter of "human physiology – feeling good"; otherwise, since the human body has not changed perceptibly in recent years, how would we explain that "our idea of what is comfortable differs from that of a hundred years ago." But has it? I have recently rented a simple log cabin in the woods north of Kingston, with no electricity, no running water, a wood-burning cookstove, kerosene lamps for lighting; it's the only house in the world I can think of that uses less nonrenewable energy or costs less to run than the Greenhome. And it is comfortable. But its comfort has nothing whatsoever to do with convenience – it is monumentally inconvenient. To turn up the heat, I have to split wood for half an hour. Keeping the kerosene lamps filled is a daily task. I have to haul water up from the lake in 5-gallon buckets. But none of that inconvenience detracts one iota from its comfort. Comfort, as Rybczynski concludes, is subjective, because "it incorporates many transparent layers of meaning, some of which are buried deeper than others." For John Kokko, comfort has to do with knowing that his home is not destroying the environment. "What am I going to say to my grandchildren?" John asked one day when we were talking about convenience. "Am I going to say, 'Sorry, I would have tried to make this a decent world for you to live in, but it just wasn't *convenient?*' "

"It may be enough," writes Rybczynski, "to realize that domestic comfort involves a range of attributes – convenience, efficiency, leisure, ease, pleasure, domesticity, intimacy and privacy – all of which contribute to the experience; common sense will do the rest."

The Greenhome has this range of attributes, I think, closing the book. Comfort, when one of its attributes is common sense, is no small thing.

SOURCES

□ □

Suggestions for Further Reading:

Cole, John N., and Charles Wing, *From the Ground Up.* Atlantic Monthly/Little, Brown, Boston, Toronto, 1976. Still the classic in commonsense owner-building.

Johnson, Lorraine, *Green Future: How to Make a World of Difference.* Penguin, Markham, Ontario, 1990. A well-researched and useful how-to for backyard environmentalists, with a foreword by Julia Langer of Friends of the Earth.

Kidder, Tracy, *House.* Houghton Mifflin, Boston, 1985.

King, Anthony D., *The Bungalow: The Production of a Global Culture.* Routledge & Kegan Paul, London, 1984.

Locke, Jim, *The Well-Built House.* Rev. ed. Houghton Mifflin, Boston, 1992. By the carpenter who built the house that Tracy Kidder wrote about.

Rybczynski, Witold, *Home: A Short History of an Idea.* Viking Penguin, New York, 1987.

– – – –, *The Most Beautiful House in the World.* Viking Penguin, New York, 1990.

– – – –, *Looking Around: A Journey Through Architecture.* Viking Penguin, New York, 1993.

– – – –, *Paper Heroes: Appropriate Technology, Panacea or Pipe Dream?* Viking Penguin, New York, 1991. "Appropriate Technology," Rybczynski writes in the foreword, "is part lay religion, part protest movement and part economic theory." An early voice in the wilderness.

Stein, Richard G., *Architecture and Energy.* Doubleday, New York, 1978. This book represents an even earlier voice: Stein was one of the first architects to consider the concept of embodied energy in building materials as the amount of energy invested in a building before it is built.

Greater Toronto Home Builders' Association, *Making a Molehill Out of a Mountain: Reducing the Volume of Residential Waste.* 1990. Available from GTHBA, 20 Upjohn Road, North York, Ontario M3B 2V9, (416) 391-3445. Price: $20.

– – – –, *Making a Molehill Out of a Mountain II: Implementing the 3 R's in Residential Construction.* 1991. An updated version of the above, which includes an emphasis on implementing solutions to the problem of construction waste. Also $20.

Wells, Malcolm, *Gentle Architecture.* McGraw-Hill, New York, 1991. A thoughtful book from one of the original low-energy architects; Wells pioneered the underground house/office building/community concept.

– – – –, editor, *Notes From the Energy Underground.* Van Nostrand Reinhold, New York, 1980. A collection of off-the-wall essays about the energy crisis.

□ □

The following is a list of sources for the low-energy materials that were used in the construction of the Greenhome. Although some of them are local, they may be able to provide suggestions as to where similar materials may be obtained in other locales. Sources for common or easy-to-find materials, such as drywall or compact fluorescent light bulbs, are not given.

☐ THE FOUNDATION AND SHELL

Recycled aggregate: Kieswetter Cartage and Excavating Co. Ltd., Box 231, Heidelberg, Ontario N0B 1Y0, (519) 699-4445. Also: Tri-City Ready Mix Ltd., Box 258, Heidelberg, Ontario N0B 1Y0, (519) 699-4880.

Recycled crushed glass for aggregate: Regional Municipality of Waterloo Waste Management Centre, 925 Erb Street West, Waterloo, Ontario N2J 3Z4, (519) 883-5118.

Precast concrete foundation: Lake Huron Precast, Box 238, Zurich, Ontario N0M 2T0, (519) 236-7670.

Polyethylene basement waterproofing (Platon System): Engineered Basement Solutions, a Division of Simon-Wood Limited, Box 188, 350 Woolwich Street South, Breslau, Ontario N0B 1M0, (519) 648-2164.

Insulating exterior sheathing: American Excelsior Co., c/o Daritek Agencies Inc., Box 434, Morin Heights, Quebec J0R 1H0, (514) 226-7968.

Cellulose insulation material: Climatizer Wall Insulation System Ltd., 120 Claireville Drive, Etobicoke, Ontario M9W 5Y3, (416) 798-1235.

Fibreglass insulation using recycled glass: Ottawa Fibre Inc., 3985 Belgreen Drive, Box 415, R.R. 4, Ottawa, Ontario K1G 3N2, (613) 736-1215.

Underslab polystyrene insulation: Plasti-Fab Ltd., 152 Birch Avenue, Box 878, Kitchener, Ontario N2G 4E1, (519) 571-1650.

Polyethylene piping for underslab cooling system: Dura-Line Corporation, Box 1445, Middlesboro, Kentucky 40965, 1-800-847-7661.

Gypsum underlayment and wallboard: Louisiana-Pacific Canada Ltd., 75 Loyalist Drive, Welland, Ontario L3C 2X9, 1-800-668-6187. In the United States: Louisiana-Pacific Corporation, 100 Northlake Drive #15, Orchard Park, New York 14127, (418) 484-1348.

Oriented-strand-board subfloor: MacMillan Bloedel Building Materials, 111 Bleams Road, Kitchener, Ontario N2C 2G2, (519) 894-2222.

Structural composite lumber and wood-I joists: Trus Joist MacMillan Limited, 86 Guided Court, Suite 10, Rexdale, Ontario M9V 4K6, 1-800-263-2325.

Steel roof: Tile Master Roofing Systems, 1060 Colborne Street East, Brantford, Ontario N3T 5M1, 1-800-461-3805.

Recycled wood siding: Canexel Hardboard Division, Canadian Pacific Forest Products Limited, 79 Main Street, Box 1020, Gatineau, Quebec J8P 6K2, (819) 643-7282.

Air-barrier house wrap: Du Pont Canada Inc., Box 2200, Streetsville, Mississauga, Ontario L5M 2H3, (416) 821-3300.

☐ WINDOWS AND DOORS

High-performance windows: Accurate Dorwin Co., 660 Nairn Avenue, Winnipeg, Manitoba R2L 0X5, (204) 667-5078.

Insulating edge spacer for windows: Edgetech I.G. Ltd., 39 Vaughan Street, Ottawa, Ontario K1M 1W9, (613) 749-0624.

Low-E glass for windows: Libbey-Owens-Ford Co., 811 Madison Avenue, Toledo, Ohio 43697-0799, (419) 247-3731.

Lineals for windows: Omniglass Ltd., Suite 9, 1329 Niakwa Road East, Winnipeg, Manitoba R2J 3T5, (204) 256-3767.

Insulated glazing unit fabrication: Sunlight Insulating Glass Manufacturing Ltd., 1416 Bonhill Road, Mississauga, Ontario L5T 1L3, (416) 564-8235.

Automatic window shading: Suntek, 6817A Academy Parkway East, Albuquerque, New Mexico 87109, (505) 345-4115.

Exterior fibreglass doors: Pease Industries Inc., 22 Noble Court, Georgetown, Ontario L7G 1M6, (416) 873-1554.

High-performance door windows: Baylite Division of Bay Mills Ltd., 7299 David Hunting Drive, Mississauga, Ontario L5S 1W3, (416) 672-2255.

Insulating spacer for door windows:

BayForm Division of Bay Mills Ltd., 500 Barmac Drive, Weston, Ontario M9L 2X8, (416) 746-0662.

☐ INTERIOR FINISHINGS

Recycled paint: Scarfe Paint Ltd., Box 217, R.R. 4, Owen Sound, Ontario N4K 5P3, (519) 376-0460.

Recycled carpet: Venture Carpets (Ontario) Ltd., 869 Gana Court, Mississauga, Ontario L5S 1N9, (416) 670-7847.

Recycled carpet underpad: Sellner Contracting Inc., 1659 Victoria Street North, Kitchener, Ontario N2B 3E6, (519) 576-7160.

☐ GAS APPLIANCES

Furnace/HRV: Canadian Gas Research Institute, 55 Scarsdale Road, Don Mills, Ontario M3B 2R3, (416) 447-6465.

Air filter: Trion Canada Inc., 130 Otonabee Drive, Kitchener, Ontario N2C 1L6, (519) 895-0570.

Fan motors: GE Motors Canada, 107 Park Street North, Peterborough, Ontario K9J 7B5, (705) 748-7183.

High-efficiency direct-vent gas water heater: Delta-Temp Corporation, 6140 Main Street, Stouffville, Ontario L4A 1A5, (416) 642-4677.

Natural gas fireplace: Powrmatic of Canada Ltd., 1155B Barmac Drive, North York, Ontario M9L 1X4, (416) 744-7206. Also: Security Chimneys Ltd., 1380 Hopkins Street, Whitby, Ontario L1N 5S1, (416) 668-8112.

Clothes dryer: Inglis Limited., 1901 Minnesota Court, Mississauga, Ontario L5N 3A7, (416) 821-6400.

Range: Canadian Gas Research Institute, 55 Scarsdale Road, Don Mills, Ontario M3B 2R3, (416) 447-6465.

Vehicle refuelling appliance: Union Gas Limited – NGV Division, 603 Kumpf Drive, Waterloo, Ontario N2V 1K8, 1-800-265-5277.

Corrugated stainless-steel piping: Hitachi, Blossburg, Pennsylvania.

□ ELECTRIC APPLIANCES AND LIGHTING

Refrigerator: Vestfrost, Spangsbjerg Mollevej, Postbox 2079, Denmark-6705 Esbjerg 0, telephone: 011-45-79-22-36.

Dishwasher, washing machine and vacuum cleaner: Miele Appliances Limited, 250 Shields Court, Unit 4, Unionville, Ontario L3R 9W7, (416) 474-1073.

Outdoor solar lighting: Siemens Solar Industries, 4650 Adohr Lane, Camarillo, California 93012, (805) 482-6800.

Occupancy sensor: Pass & Seymour Canada Inc., 448 North Rivermede Road, Concord, Ontario L4K 3M9, (416) 738-9195.

□ PLUMBING

2-litre flush toilet: Control Fluidics, Inc., 124 West Putnam Avenue, Greenwich, Connecticut 06830, (203) 661-5599.

3-litre flush toilet: Water Conservation Systems, Inc., Damonmill Square, Nine Pond Lane, Concord, Massachusetts 01742, (508) 369-6037.

Shower head and faucets: Moen Inc., 2816 Bristol Circle, Oakville, Ontario L6H 5S7, 1-800-465-0279.

Automatic infrared faucet: Nepitek Ltd., Box 5581, Station F, Nepean, Ontario K2C 3M1, (613) 723-8090.

Photovoltaic-powered solar water heater: Thermo Dynamics Ltd., 81 Thornhill Drive, Dartmouth, Nova Scotia B3B 1R9, (902) 468-1001.

□ □

The Greenhome in Waterloo is one of 10 houses being built under the Department of Energy, Mines and Resources' Advanced Houses program. The locations and contact persons for all of them are given here. Each house incorporates advanced construction techniques and energy-efficiency features, and all are well worth a visit.

1. The Waterloo Greenhome, in Ontario. Contact: Stephen Carpenter, Enermodal Engineering Ltd., (519) 884-6421.

2. The P.E.I. Advanced House, near Charlottetown. Contact: Norman Finlayson, P.E.I. R-2000 Office, (902) 368-3303.

3. The Neat Home, in Hamilton, Ontario. Contact: Don Buchan, Buchan, Lawton, Parent, (613) 748-3762.

4. Maison Performante, in Laval, Quebec. Contact: Hugh Ward, l'Association provinciale des constructeurs

d'habitations du Québec, (514) 353-1120.

5. The Envirohome, in Bedford, Nova Scotia. Contact: Dick Miller, Clayton Developments, (902) 445-2000.

6. Saskatchewan Advanced Technology House, in Saskatoon. Contact: John Carroll, Carroll Homes, (306) 955-6677.

7. The British Columbia Advanced House, in Surrey. Contact: Richard Kadulski, Richard Kadulski Architect, (604) 689-1841.

8. The Manitoba Advanced House, in South Winnipeg. Contact: Don Glays, Manitoba Home Builders' Association, (204) 477-5110.

9. Innova House, in Kanata, Ontario. Contact: Bruce Gough, Energy Building Group, (613) 723-5907.

10. Maison Novtec, in Montreal, Quebec. Contact: Dr. Paul Fazio, (514) 848-8770.

For more information about the Advanced Houses program, contact:

CANMET Buildings Group, Energy, Mines and Resources Canada, 580 Booth Street, Ottawa, Ontario K1A 0E4, or Canadian Home Builders' Association, 150 Laurier Avenue West, Suite 200, Ottawa, Ontario K1P 5J4.

INDEX

Advanced Houses program
  objectives, 46-52
  philosophy, 12-13
ADW house design, 1983, 33-34
Air quality, indoor, 176-177
Air-to-air heat exchangers. See Heat-
  recovery ventilators
Air/vapour barriers, 122-124
Airtight construction, 26
  in Advanced Houses, 47
Aluminum siding, 168-169
Appliances, 141-152
  energy consumption, 51, 143, 144
  Greenhome's, 144-151
Argisol, 80-81
Argon-filled windows, 116
Aspen, 105-106
Basements, 68-73, 74-75
  Greenhome's, 73, 74-75; floor plan,
  71
  questionable need for, 69-70
Blue Box program, 170-172
Brampton House, 38-39, 43
Building materials
  environmental assessment of, 48,
  49
  sources, 201-204
  waste, 59, 60, 99
Building practices, traditional,
  problems with, 56-57
Bungalows, 96-99
Carpeting, recycled, 173-174
Cellulose insulation, 120-122
Cementatious wood, 81-82
CFCs and ozone, 119-120
Chlorofluorocarbons and ozone,
  119-120
Cisterns, 73-74
  Greenhome's, 74, 187
Clothes washers, 148-150
Compact fluorescent lamps, 139
Computer predictions of energy use,
  29-30, 41-42
Concrete
  blocks, 75-77
  poured, 75-76, 77

precast, 82, 83-84, 87
Construction practices
  environmental impact of, 64
  traditional, problems with, 56-57
  waste, 59, 60, 99
Cosmetic concerns with low-energy
  house design, 29
Decking, 105
Dishwashers, 146-147
Doors, 112, 113-114
Drainage material, 90
Dryers, 150-151
Durisol Wall-Forms, 80
ECMs, 155
Education on low-energy house
  design, 13, 30, 38, 43-44
Electric heating, 129
Electrical wiring, installation, 152
Electricity
  generation, environmental impact,
  60-61
  generation, wastefulness of, 129
  overproduction, 128-129
  value of, 133
  waste of, 58, 60-61
Electronically commutated motors,
  155
EnerGuide program, 143
Energy
  embodied, 60
  renewable. See Solar energy; Solar
  heating
Energy consumption of appliances,
  51, 143, 144
Energy crisis, 1973, 13-15
Energy efficiency of Advanced
  Houses, 46, 49-50
Energy Showcase Project,
  Saskatchewan, 29-30
Energy use
  computer predictions, 29-30, 41-42
  Greenhome's compared with
  average house, 8-9
Energy waste, 58, 60-61
  in generating electricity, 129
EnergyGuide program, 143

Enermodal Engineering Ltd., 40-41
Enerpass, 41
Environmental costs of wasting
    energy, 61
Environmental impact of Advanced
    Houses, 51-52
Environmental impact of generating
    electricity, 60-61
Environmental impact of house
    construction, 64
Excavating the Greenhome's
    basement, 44-46
Excel Board, 122
Faucets, 187-188
Fibreglass insulation, 118-119
Fireplace, 161
Flair subdivision, Winnipeg, 32
Floor joists, 102-104; illus., 102
Flooring
    hardwood, recycled, 178
    tiles, 177-178
Floors, 108-109
Fluorescent lamps, 137-139
    compact, 139
Footings, 85
Form-a-Cell, 120
Foundations
    Greenhome's, 82-84, 99-100; illus.,
        83, 86
    materials for, 75-82
FRAME, 41
Freon, 119
Fuel, substitutes for oil, 15-16
Funding for the Greenhome, 53-55
Furnace ducts, 153
Furnaces, natural-gas, 158-159
Gas. See Natural-gas line installation
Glass, crushed, as drainage material,
    90
Greenhome, basic design, 63
Ground covers, 189-191
Groundwater depletion, 181
HCFCs, 120
Heat-recovery ventilators, 156-161
Heating ducts, 153
Heating systems. See also Electric

heating; Furnaces; Heat-recovery
    ventilators; Solar heating
efficiency in Advanced Houses, 52
Greenhome's, 153-161; illus., 154
Heating vents, positioning, 153-156
Honeywell TotalHome System, 133
Hot-water heaters, 148
    Greenhome's, illus., 149
House size, 93-95
    environmental impact of, 98-99
HRVs, 156-161
HSLOAD, 30
Hydrochlorofluorocarbons, 120
IMS, 33-34, 158-159
Incandescent lamps, 137
Indoor air quality, 176-177
Insulation, 22, 23-24, 118-122
    Greenhome's, 120-122; illus., 121
    and solar heating, 22, 23-24
    superinsulation, 26
    window, 27-28
Integrated heating-and-ventilating
    appliance, 161
Integrated mechanical system, 33-34,
    158-159
Landscaping, 188-192; illus., 190
Lawn, 189
Life-cycle costing, 151-152
Light, colours of, 138
Light-and-tight approach, 25-26
Lighting, 134-140
    Greenhome's, 139-140
    levels, recommended, 135
    requirements, 136-137
Low-E windows, 115
Low-energy house design
    cosmetic concerns, 29
    history, 17-34, 36-39, 41-44
    philosophy, 42-43, 64, 197
    promoting knowledge of, 13, 30,
        38, 43-44
Main-floor plan, 95
Mass-and-glass approach, 25
    windows, 26-27
Mineral-fibre insulation, 119
Moisture barrier, 87-90; illus., 88

Natural-gas line installation, 127-128
Oil
  depletion of resources, 13-15
  substitutes for, 15-16
Ontario Hydro, 126-127, 128-129
Ozone and CFCs, 119-120
Ozone-depletion factor, 51-52
Paint, recycled, 174-176
Parallel-strand lumber, 106
Photovoltaic cell, discovery of, 17
Platon system, 87-90; illus., 88
Preserved-wood foundations, 77-80
R-2000 program, 30-33, 42-43
  disadvantages of, 32-33
Radon, 89
Ranges, 147-148
Reading materials, 200
Recycling, 170-173, 178-180
  facilities in Advanced Houses, 51
Red Box program, 175
Refrigerators, 143, 144-146
Roofing materials, 164-167
Roofs, 109-112
  trusses, 110-112
Saskatchewan Conservation House,
  19-21; illus., 20
Shakes, wood, 164
Shingles
  asphalt, 164-165
  recycled-wood, 167
  wood, 164
Showers, 187
Siding, 167-170
  aluminum, 168-169
  recycled-wood, 167-168
  vinyl, 169-170
Smart Houses, 130-133
Solar energy, 16-24
Solar heating, 17-24
  active, 17-19
  and insulation, 22, 23-24
  passive, 17, 21-22
  passive, light-and-tight approach,
    25-26
  passive, mass-and-glass approach,
    25

  and windows, 25-28
Sources, 200-204
SpaceJoist system, 103-104
Spacers for windows, 116, 117
Sparfil, 80
Steel roofing, 166-167
Stoves, 147-148
Styrofoam SM insulation, 119-120
Surveying the Greenhome lot, 10-12
Toilets, 182-183, 186-187
TRON House, 131
Trus Joist Silent Floor, 102
Ventilation systems for Advanced
  Houses, 47-49
Venting of combustion equipment,
  49
Vinyl siding, 169-170
Walls
  exterior, 106-108; illus., 101, 107
  exterior, sheathing, 122
  interior, 109
Washing machines, 148-150
Waste
  construction, 59, 60
  energy, 58, 60-61
  reduction, 179
Water systems, Greenhome's, 74;
  illus., 72
Water usage, 180-188
  in Advanced Houses, 50-51
  Greenhome's, 184-188
Weeping-tile bed, 85
Windows, 25-28, 114-118
  automatic shading of, 195
  computer predictions of
    performance, 41
  controlling heat loss, 25-28
  energy ratings, 116
  Greenhome's, 116-118; illus., 117
  improvements to, 28
  insulation for, 27-28
Wood
  aspen, 105-106
  recycled, shingles, 167
  recycled, siding, 167-168
  usage, reduction of, 103, 112